# *SILAGE FOR MILK PRODUCTION*

EDITORS:— J.A.F. Rook & P.C. Thomas

**Technical Bulletin 2**

National Institute for Research in Dairying, Reading, England
Hannah Research Institute, Ayr, Scotland

*First published......................................1982*



Technical Bulletin Editors
J.A. Irvine & P.D. Wilson


ISBN 0-7084-0166-X

For titles of other publications in the series
and for ordering, see inside back cover

Printed at the College of Estate Management, Reading

# Contents

## Foreword

From its foundation, the Hannah Research Institute has been involved in research into the production of grass and its use in grazed or conserved form by the dairy cow. For much of the period emphasis has been given to forage quality, and to low-cost milk production from home-grown forage foods. In recent years, in particular, attention has been focused on grass silage, which appears to offer a technically and economically acceptable means of providing high-quality fodder for the winter period. The work on silage is continuing: new findings will emerge and new developments will take place. However, the work that has been conducted already has led to a well tried and tested procedure for the production of high-quality silages on a farm scale, and has demonstrated that these silages can make a major contribution to the winter feeding of the dairy cow.

This bulletin is intended to give an insight into grassland management and into the principles of silage making, and to summarize the results and practical experience gained in the experimental studies conducted at the Institute. The information provided is directed towards readers wishing to advise on or to make high-quality silage and to exploit the potential of such silage as a food for dairy cows.

We are most grateful to the authors and to Mrs J. Briggs, Mr B.F. Bone, Mrs J. Geens, Mr J.A. Irvine and Dr P.D. Wilson, who were involved in the preparation of the bulletin.

J.A.F. Rook & P.C. Thomas
*November 1981*

# Chapter 1

## Introduction

**J.A.F. Rook**

*Hannah Research Institute, Ayr KA6 5HL*†

Work with silage was undertaken at the Hannah Research Institute as early as 1936, when a comparison was made of the nutritive value for milk production of the proteins of grass ensiled without treatment or after the addition of molasses or mineral acids (Morris *et al.*, 1936). At a later date, as part of a wider study aimed at improving the contribution of grassland to milk production, an assessment was made of the value of silage as a component of mixed forage diets (Reid & Holmes, 1956; Holmes *et al.*, 1957). Following the introduction of the forage harvester and reliable methods of applying silage additives in the mid 1960s, however, experiments were begun in which silage was the main forage component of the diet and where the conservation of grass as silage was seen as an integral part of the use of grassland for milk production.

The early trials demonstrated a highly significant positive relationship between the digestibility of silage offered *ad lib.* and milk yield. With 11 grass silages that ranged in digestibility from 600 to 700 g digestible organic matter per kg dry matter the mean daily milk yield was increased by 0.20 kg with each 10 g. $kg^{-1}$ increase in digestibility. The improvement in milk yield was associated with a higher intake of both dry matter and digestible organic matter of the more highly digestible silages (Castle, 1975).

It was recognized that the production of silage of high digestibility required ensilage of grass at an early stage of growth, and that harvesting at that stage might reduce the annual yield of nutrients. However, there appeared to be a need for the development of a low-cost system of milk production using relatively inexpensive home-grown foods and low levels of purchased concentrates, because of likely future adverse changes in the economics of milk production, and it seemed possible that high-quality silage might be an especially valuable component of such a system.

Feeding costs account for more than half of the total costs of milk production and, with conventional systems of feeding, purchased foods

† Present address: Agricultural Research Council, 160 Great Portland Street, London W1N 6DT

contribute two-thirds of the feeding costs. Although a high use of concentrated foods was at that time, and still is, justified by the relative prices for concentrates and milk, a further narrowing of the price differential seemed, and still seems, inevitable. The substantial expansion of milk production since 1950, when 0.90 of that produced was used for liquid consumption, has meant that an increasing proportion of milk has been used for manufacture; the proportion is now approximately half. The market return for milk used for the production of traditional dairy products is less than for that used for liquid consumption, and the overall return could be depressed further because of the current surplus of milk within the European Economic Community and by any fall in the consumption of liquid milk. There is also the prospect that, with the continuing expansion of the world population, cereal use might be diverted away from animal feeding towards consumption by man and that, relatively, cereal prices will rise. The production of milk at lower cost would, at a given market price, provide a better margin per unit of production and protect the individual farmer against an enforced cut in production. Alternatively, a fall in the market price of milk would make it more competitive with other new materials and improve marketing opportunities.

For low-cost, high-forage systems to be viable, however, production must be maintained at a high enough level to provide satisfactory returns on capital investment and other costs, and that requires adequately high yields of milk and, consequently the provision of diets with a high metabolizable energy concentration. With well managed swards, these criteria may be readily met for much of the grazing season but during the winter period the provision of conserved forage of high quality is essential. Silage of high digestibility appeared to offer a technically and economically feasible means of meeting the requirement.

In 1973, therefore, the decision was taken to investigate aspects of the production and use of high-quality silage — the composition and management of the sward, the technique of ensilage and the system of feeding — to allow an assessment of its nutritive value, and to identify technical and nutritional limitations on its production and use for milk production. This bulletin summarizes results obtained over the following 6 years and places them in context with work conducted at other research centres. The bulletin is not intended as a textbook on grass silage nor as a comprehensive review, but it was considered that an appreciation of important features of grass production and of the ensilage process would provide the necessary theoretical background to the more specific consideration of the Institute's work on high-quality silage. The first part of the bulletin, therefore, deals with the chemistry of silage making, with selected aspects of the composition and management of grass swards, and with factors influencing the nutritive value of silages, whilst the second part concentrates on the making and feeding of high-quality silage. Inevitably, the individual chapters present technical information at different levels and, whereas the first part is concerned largely with the principles of grass production and conservation, the second part deals with the practical details and implications of a particular well tried conservation procedure, and with possible future developments. The two parts will not have equal appeal to all readers but it is hoped that advisers, students and farmers will find something of value in both.

## REFERENCES

CASTLE, M.E. 1975 Silage and milk production. *Agricultural Progress* **50** 53-60

HOLMES, W., REID, D., MACLUSKY, D.S., WAITE, R. & WATSON, J.N. 1957 Winter feeding of dairy cows. IV. The influence of four levels of concentrate feeding in addition to a basal ration of grass products on the production obtained from milking cows. *Journal of Dairy Research* **24** 1-10

MORRIS, S., WRIGHT, N.C. & FOWLER, A.B. 1936 The nutritive value of proteins for milk production. IV. A comparison of the proteins of (*a*) spring and autumn grass, (*b*) grass conserved as silage (A.I.V. acid treated, molasses treated and ordinary untreated), and (*c*) grass conserved by drying, with notes on (i) the effect of heat treatment on the nutritive value and (ii) the supplementary relations of food proteins. *Journal of Dairy Research* **7** 97-121

REID, D. & HOLMES, W. 1956 Winter feeding of dairy cows. III. The influence on milk yield of high and low protein concentrates, each fed at two levels in addition to dried grass and grass silage. *Journal of Dairy Research* **23** 159-168

# CONSERVATION OF HERBAGE AS SILAGE

# Chapter 2

## The chemistry of silage making

**P.C. Thomas & I.M. Morrison**

*Hannah Research Institute, Ayr KA6 5HL*

*Ensilage* may be used as a general term to describe any procedure involving the storage of materials in silos or pits. Commonly, however, the term is used specifically to describe the storage of green fodders under anaerobic conditions that allow naturally occurring microbes to ferment plant carbohydrates to organic acids, reducing the pH in the silo, inhibiting further fermentation and preserving the crop as *silage*.

### A BRIEF HISTORY

The origins of ensilage as a method of forage conservation are virtually lost in antiquity but there is evidence that the procedure was in use in Egypt as long ago as 1000-1500 BC (see Schukking, 1976). However, interest in the technique of ensilage began in earnest in the 1800s with the publication in France of Goffart's work on the ensilage of maize and other green crops (Goffart, 1877). The ensilage method he recommended involved rapid filling of the silo, with compaction and tight sealing of the surface to exclude air, and in many instances this probably gave 'low-temperature' (below 30°C) fermentation conditions not unlike those advocated today.

Procedures similar to Goffart's had, in fact, been described earlier in Scotland for the ensilage of grass (Johnston, 1843) but the impetus to silage making in Britain came rather later with the publication of a book on ensilage by Fry (1885). Differing from earlier workers, Fry advocated that the rate of filling and degree of compaction of the crop in the silo should be adjusted to encourage initial heating to raise the temperature above 50°C. This procedure, which seems to have been widely adopted in practice, was designed to produce 'sweet silages', in contrast to the 'sour' or 'acid' silages produced by other methods.

Fry's ideas were far from universally accepted nonetheless, and a number of researchers were quick to point out that heating in the silo could not occur without loss of nutrients. Throughout the early 1900s, especially on the continent, work on low-temperature methods of ensilage continued and attention was focused on the chemical, physical and microbiological factors that influenced the fermentation characteristics. This led to major advances in silage making between 1925 and 1939. Individual developments during this period are too numerous to give details but two events warrant special mention. In Finland, A.I. Virtanen began work on silage additives, which led to the introduction of additives containing inorganic and, later, organic acids (Virtanen, 1938), and, in Britain, S.J. Watson began studies that were to form the basis for his books (Watson, 1938 & 1939), which were to have an important and lasting impact on ideas about ensilage and conservation throughout the country.

At the time his books were published, Watson was of the opinion that the preferred ensilage procedure should involve a warm fermentation, intermediate between the two extremes previously advocated. The recommendation was that the rate of filling and degree of compaction in the silo should be adjusted to allow limited heating to approximately 37°C, to encourage active microbial fermentation whilst avoiding over-

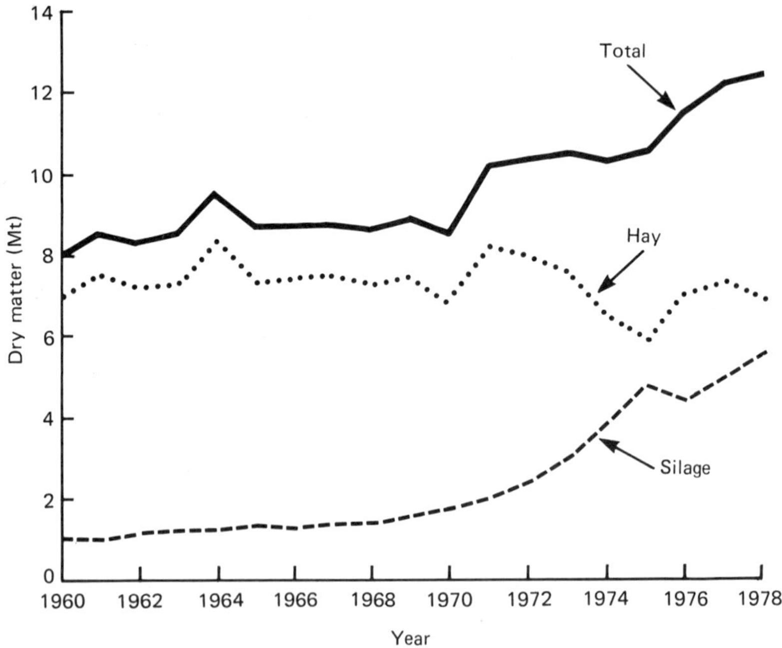


**Fig.1. Annual production of conserved grass products in the United Kingdom from 1960 to 1978 (from Wilkinson, 1981)**

heating and crop damage. The approach became popular on British farms although, under conditions where farm silos could be unlined pits dug out of the ground, ensilage was something of a gamble with unpredictable and sometimes unacceptable results. The practical problems of silage making were, however, greatly reduced during the 1950s and 1960s as a result of a number of innovations. These included flail- and later precision-chop forage harvesters, which improved the collection and pre-ensiling treatment of grass, better designed and constructed silos, polyethylene sheeting to exclude air, and effective silage additives. Progressively there has been a movement towards 'low-temperature' ensilage procedures, with more predictable and consistent results. Reflecting this, the amount of silage made on British farms has increased sharply since the early 1970s (fig. 1) and for many farmers ensilage is now the preferred method of forage conservation.

## PRINCIPLES OF CONSERVATION

There is a wide choice in the detail of the procedures that may be used for silage making but the success and efficiency of the conservation process, and the feeding value of the silage produced, depend on a limited number of factors. Chief amongst these are the chemical composition of the crop, the post-mowing, pre-ensiling changes in crop composition, and the fermentation in the silo.

### The composition of the crop

Water is the most abundant constituent in all green forages. It is held within the plant cells, partly bound to cellular materials, and partly free and acting as a solvent for water-soluble constituents. The water content of forages varies with the species and strain of plant, with the stage of growth, with fertilizer treatment and with climatic environment but, for leafy materials typical of those used for silage making, water usually accounts for 780-850 g. $kg^{-1}$ of the total weight of the crop. The remainder consists of carbohydrates, lipids, nitrogenous constituents and minerals (table 1).

The plant carbohydrates may be classified into two main groups: the water-soluble or non-structural carbohydrates that are found in the plant cells, and the structural carbohydrates that are components of plant cell walls. The water-soluble carbohydrates include simple sugars (mainly glucose, fructose and sucrose) and organic acid intermediates of sugar

**Table 1. Typical chemical composition of grasses and legumes (g. $kg^{-1}$ dry matter)†**

| | Grasses | Legumes |
|---|---|---|
| Carbohydrates | | |
| Crude fibre | 180-260 | 140-220 |
| Nitrogen-free extract | 470-600 | 350-420 |
| Lipids | 10-50 | 20-40 |
| Crude protein | 130-240 | 220-330 |
| Minerals | 70-100 | 90-120 |

† Values refer to primary growths cut in the spring at a conventional silage stage and are based on results for ryegrass, timothy, cocksfoot and tall fescue, and white clover, red clover and lucerne. For a survey of the chemical composition of forages see Agricultural Research Council (1976)

metabolism (malate, citrate, oxaloacetate etc.), along with complex carbohydrates (polysaccharides) that the plant synthesizes as an energy store (table 2). In temperate grasses the energy store is in the form of polysaccharides called fructosans, which are made up of fructose units.

**Table 2. The concentrations of sugars and storage polysaccharides in some samples of grasses and legumes (g. $kg^{-1}$ dry matter)†**

| | Glucose+ fructose | Sucrose | Fructosan | Starch |
|---|---|---|---|---|
| Perennial ryegrass | 46 | 32 | 60 | - |
| Cocksfoot | 36 | 48 | 36 | - |
| Red clover | 47 | 48 | - | 86 |
| Lucerne | 50 | 42 | - | 87 |

† Values are for leafy material at an immature stage of growth

**Table 3. The concentrations of cell wall components in samples of ryegrass and lucerne (g. $kg^{-1}$ dry matter)†**

| | Perennial ryegrass | Lucerne |
|---|---|---|
| Cellulose | 213 | 118 |
| Hemicellulose | 158 | 123 |
| Pectin | 24 | 123 |
| Lignin | 27 | 64 |

† Values are for leafy material at an immature stage of growth

In legumes the predominant storage polysaccharide is starch, which is made up of glucose units. The cell wall is often referred to as the plant 'fibre' although this is not a discrete chemical compound but an association of a number of components. One of the carbohydrate components is cellulose (table 3) which, like starch, is made up of glucose units but which differs from starch in the way that the units are linked together (cellulose has $\beta$ 1-4 bonds whilst starch has $\alpha$ 1-4 and $\alpha$ 1-6 bonds). The cellulose chains are simple linear molecules, but in the plant many chains may be aligned and cross-linked by hydrogen bonds to form a highly ordered molecular complex. In addition, the cell wall contains a range of other carbohydrate and non-carbohydrate components, and current views are that these materials are also linked together in a complex by covalent bonds. The main carbohydrate in this complex is hemicellulose which, in reality, is not a single chemical compound like cellulose but a family of compounds. Thus xylans are hemicelluloses made up of the sugar xylose whilst mannans, galactans and glucans are hemicelluloses composed of the sugars mannose, galactose and glucose respectively. Few hemicelluloses actually contain a single sugar and the main hemicelluloses in grasses are arabinoxylans (i.e. made up of arabinose and xylose), whilst in legumes they are arabinoxylans and glucomannans (glucose and mannose). Legumes and, to a lesser extent, grasses additionally contain the polysaccharide pectin which is made up of sugars and acidic sugar derivatives, e.g. galacturonic acid. The major non-carbohydrate component in the cell wall complex is the aromatic polymer lignin, but there are small amounts of phenolic acids and acetic acid that are esterified to the other fibre components.

Plant lipids are a mixture of triglycerides, glycolipids, phospholipids, waxes and sterols. The first three of these components in particular provide a significant dietary source of long-chain fatty acids for the

**Table 4. Fatty acid composition of early growth and mature ryegrass (g. kg$^{-1}$) (after Hawke, 1973)**

| Fatty acid | Abbreviation | Early growth | Mature growth |
|---|---|---|---|
| Lauric acid | 12:0 | 2 | 5 |
| Myristic acid | 14:0 | 6 | 10 |
| Palmitic acid | 16:0 | 112 | 169 |
| Palmitoleic acid | 16:1 | 16 | 19 |
| Stearic acid | 18:0 | 7 | 10 |
| Oleic acid | 18:1 | 14 | 27 |
| Linoleic acid | 18:2 | 84 | 127 |
| Linolenic acid | 18:3 | 760 | 632 |

animal and, characteristically, the mixture of fatty acids is rich in unsaturated acids (table 4).

The forage nitrogenous components include proteins and non-protein nitrogenous materials such as nitrates, ammonium salts, peptides, free amino acids, and the purine and pyrimidine bases of nucleic acids. Generally, 750-850 g. $kg^{-1}$ of the total nitrogen is accounted for as plant protein, the amino acid composition of which appears to vary relatively little with plant source (table 5).

**Table 5. The amino acid composition of some samples of forage protein (g amino acid per kg total amino acid) (based on data from Lyttleton, 1973)**

| Amino acid | Ryegrass | Cocksfoot | Lucerne |
|---|---|---|---|
| Aspartic acid | 103.9 | 102.3 | 106.0 |
| Threonine | 50.4 | 52.4 | 51.5 |
| Serine | 36.6 | 39.9 | 40.1 |
| Glutamic acid | 136.8 | 134.7 | 125.9 |
| Proline | 49.7 | 55.7 | 51.7 |
| Glycine | 34.3 | 38.3 | 36.3 |
| Alanine | 51.0 | 50.8 | 46.1 |
| Valine | 62.2 | 62.7 | 67.5 |
| Methionine | 26.0 | 28.4 | 22.7 |
| Isoleucine | 56.7 | 56.7 | 64.8 |
| Leucine | 99.3 | 103.3 | 104.7 |
| Tyrosine | 67.6 | 57.9 | 67.6 |
| Phenylalanine | 83.6 | 86.6 | 85.4 |
| Histidine | 12.0 | 11.2 | 11.4 |
| Lysine | 80.4 | 79.8 | 82.0 |
| Arginine | 25.0 | 24.5 | 21.8 |
| Cystine | 9.0 | 14.6 | 14.2 |
| Tryptophan | 15.3 | - | - |

## *Factors influencing crop composition*

The chemical composition of forage crops is influenced by a large number of factors including the type, species and strain of plant, growing conditions (e.g. temperature, light intensity and water availability), fertilizer treatment, stage of growth, and cutting or grazing management. Moreover, the factors are subject to complex interactions, partly because of differential effects on the relative rates of growth and maturation of the plant's leaf blade, leaf sheath, stem and inflorescence. Even for a single strain of grass or clover it is impossible to predict the chemical composition under a particular set of field conditions with precision (see Agricultural Research Council, 1976). Some important and consistent trends in composition should, however, be noted. Amongst

**Table 6. The chemical composition of ryegrass cut at differing stages of growth (g. kg$^{-1}$ dry matter). Data from Waite *et al.* (1964)†**

| | Young leafy | | Late leafy | | Head emergence | | Seed setting | |
|---|---|---|---|---|---|---|---|---|
| Protein nitrogen x 6.25 | 146 | 185 | 116 | 152 | 106 | 138 | 64 | 96 |
| Non-protein N x 6.25 | 39 | | 36 | | 32 | | 32 | |
| Lipid | | 91 | | 76 | | 65 | | 47 |
| Hexoses | 46 | | 63 | | 67 | | 40 | |
| Sucrose | 32 | 138 | 27 | 118 | 28 | 113 | 28 | 106 |
| Fructosan | 60 | | 28 | | 18 | | 38 | |
| Organic acids | | 42 | | 49 | | 46 | | 29 |
| Cellulose | | 213 | | 221 | | 239 | | 267 |
| Hemicellulose comprising | | | | | | | | |
| Xylan | 69 | | 89 | | 101 | | 144 | |
| Araban | 21 | | 22 | | 23 | | 25 | |
| Glucan | 19 | 158 | 22 | 189 | 20 | 194 | 17 | 257 |
| Galactan | 7 | | 7 | | 7 | | 7 | |
| Acidic sugars | 42 | | 49 | | 43 | | 64 | |
| Pectin | | 24 | | 21 | | 22 | | 22 |
| Lignin | | 27 | | 36 | | 43 | | 73 |
| Ash | | 81 | | 85 | | 78 | | 58 |

† Samples are of artificially dried grass made from S23 ryegrass.

these are the changes that occur as plants mature, which are associated with a reduction in the proportion of leaf, and alterations in leaf and stem composition. Characteristically, the contents of water-soluble carbohydrates, protein and lipid are reduced and the contents of cellulose, hemicellulose and lignin are increased (table 6). Lignin is not digested by the cow, and its bonding to hemicellulose and the 'encrustation' of the cellulose 'fibres' with the lignin-hemicellulose complex also limits the extent to which these carbohydrates are digested. Therefore, with advancing maturity and increased lignification, the digestibility of the cell wall components is reduced and there are accompanying effects on the digestibility of organic matter, gross energy and protein.

Low contents of water-soluble carbohydrates are also found in samples of regrowth grass and in these circumstances they are accompanied by high protein contents (fig. 2). Also, in both primary and regrowth cuts, water-soluble carbohydrate content is reduced by the application of nitrogenous fertilizers. The effect of such fertilizer on crude protein content is more complicated; when the fertilizer is applied 1-2 weeks before cutting, protein content may be elevated but, when the nitrogen is applied, as is more typical, soon after each cut, the protein content of the next cut may be unchanged or reduced (see ARC, 1976).

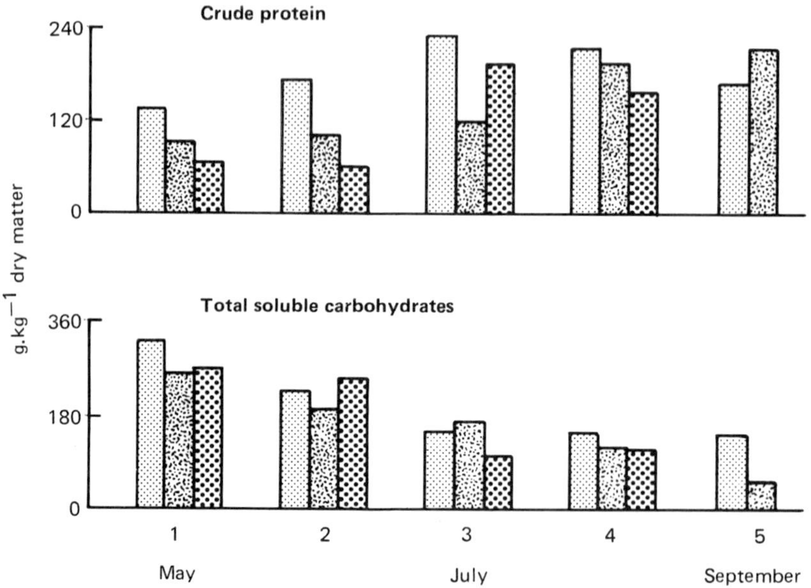


**Fig.2. Water-soluble carbohydrate and crude protein contents in cuts of three varieties of ryegrass taken each time the sward reached grazing height (after Waite, 1965)** ( ▫ Italian ryegrass; ▫ S24 ryegrass; ▫ S23 ryegrass)

## *Post-mowing changes in crop composition*

Growing plants have a high rate of transpiration through their stomata, the small openings in the surface of the leaves and stems. After the plants have been cut there is no longer a supply of water from the roots and, since water loss continues, the plant cells begin to dehydrate, lose their turgidity and wilt. Normal transpiration is inhibited but water is still lost through the stomata and to a lesser degree through the cuticle layer. The rate of loss under these conditions depends on the difference between the vapour pressure of the internal water near the surface of the plant cell and the vapour pressure of the water in the surrounding air (Sullivan, 1973). The effective value of the latter depends in turn on the temperature, humidity and movement of air in contact with the cut crop and on the addition of surface water as dew or rain, where it occurs. Water loss has been measured under a wide range of weather and management conditions, and average rates of 0.5-1.0% of the water per h have been observed in a number of studies with undisturbed swathes. These rates may be increased to approximately 2.0% per h through tedding and, if the herbage has been mechanically damaged by flailing, bruising or lacerating during cutting, rates of 3.0% per h or greater may be achieved under favourable weather conditions.

During the wilting period, changes in chemical composition are brought about by the continuing action of the plant hydrolytic and respiratory enzymes. There is hydrolysis of sucrose, fructosans and possibly starch to their constituent sugars, and oxidation of the sugars and of organic acids to carbon dioxide and water. The detailed changes in plant composition observed vary somewhat with the conditions, and with the relative rates of hydrolysis and oxidation, but under all circumstances there is a loss of total soluble carbohydrate, which is reflected in a slightly reduced dry matter yield. Reductions in soluble carbohydrate content of approximately 20% of the total content have been reported (Wylam, 1953), although it should be remembered that, despite this, the loss of water during wilting normally ensures that the soluble carbohydrate content of the fresh material is increased.

Protease enzymes are also active in the cut plant and there is a progressive hydrolysis of plant proteins to simpler non-protein nitrogenous compounds such as peptides, amino acids, amines and ammonia. A little loss of nitrogen may occur but the main effect is an alteration in the plant's composition. As with carbohydrate breakdown the extent of the effects on protein depend on the enzymic activity and the period of wilting but over the first 24 h the changes in composition are substantial (table 7). Moreover, the amino acid composition of the non-protein fraction is not the same as that of the parent protein. Most amino acids, e.g. glycine, serine, threonine, alanine, tyrosine, valine, methionine, leucine and isoleucine, are present in lowered concentration whilst the concentration of some, e.g. proline, glutamine and asparagine, is increased (Kemble & Macpherson, 1954).

## Silage fermentation

The objective in silage making is to achieve preservation whilst minimizing losses of nutrients and avoiding adverse changes in the chemical

Table 7. Nitrogen distribution in ryegrass wilted for different periods of time (from Brady, 1960)

| Wilting time (h) | Dry matter (g. $kg^{-1}$) | Total nitrogen (g. $kg^{-1}$ dry matter) | Nitrogen (g. $kg^{-1}$ total nitrogen) | | | | |
|---|---|---|---|---|---|---|---|
| | | | Total soluble | Soluble non-protein nitrogen | Amino-nitrogen | Volatile nitrogen | Amide-nitrogen |
| 0 | 161 | 202 | 170 | 89 | 26 | 1 | 8 |
| 2.5 | 192 | 183 | 164 | 114 | 59 | 2 | 8 |
| 6.5 | 268 | 173 | 171 | 127 | 71 | 2 | 7 |
| 26.5 | 390 | 179 | 189 | 173 | 92 | 6 | 13 |

**Table 8. Some species of lactic acid bacteria commonly found on fresh herbage and in silage (McDonald, 1976)**

| Homofermentative species | Heterofermentative species |
|---|---|
| *Lactobacillus plantarum* | *Lactobacillus brevis* |
| *Pediococcus acidilactici* | *Lactobacillus buchneri* |
| *Streptococcus durans* | *Lactobacillus fermentum* |
| *Streptococcus faecalis* | *Lactobacillus viridescens* |
| *Streptococcus faecium* | *Leuçonostoc mesenteroides* |
| *Streptococcus lactis* | |

composition of the crop. This is achieved through enclosure of the crop in the silo under anaerobic conditions, which allow rapid proliferation of lactic acid bacteria that produce lactic acid alone (homofermentative types) or lactic acid in mixture with other products (heterofermentative types). The end-products of the fermentation of soluble sugars by these organisms, principally lactic acid, reduce the pH in the silo until a value of approximately 4 (typically 3.8-4.2) is achieved. Then fermentation ceases and a stable preserved silage is obtained.

A large number of species of lactic acid bacteria which are present in silage have been isolated and identified (table 8) and the biochemical pathways through which they metabolize simple hexose and pentose sugars have been described (fig. 3). The organisms will also metabolize the plant organic acids such as citrate, malate and oxaloacetate to lactate, acetate, ethanol and 2,3-butanediol, and undertake partial breakdown (deamination or decarboxylation) of certain amino acids. Thus loss of amino groups with liberation of ammonia may lead to the formation of citrulline and ornithine from arginine, glutamate from glutamine, aspartate from asparagine and pyruvate from serine. Correspondingly, decarboxylation reactions, involving a loss of carbon dioxide, yield amines: tyramine from tyrosine, histamine from histidine, cadaverine from lysine and putrescine from ornithine.

In addition to the lactic acid bacteria other organisms, yeasts, fungi and especially clostridial bacteria, are present in the silo and if these organisms proliferate the pattern of silage fermentation is adversely affected. The clostridia metabolize sugars and lactic acid to produce butyrate (fig.4) and a range of lesser products including acetate, propionate, ethanol, butanol and formate. They also degrade amino acids extensively to produce a wide range of fermentation end-products (table 9). Since butyric acid is a weaker acid than lactic acid, and many of the nitrogenous products of amino acid breakdown are bases, clostridial activity slows or reverses the normal reduction in pH which occurs in the silo. The period of fermentation is prolonged and there is spoilage of the crop, with a resultant silage of low nutritive value and poor acceptability to animals.

Uncut grass carries low numbers of lactic acid bacteria, mainly on the bruised or dead parts of the plants; clostridia are also present although only in the endospore form. Following harvesting, the numbers of lactic

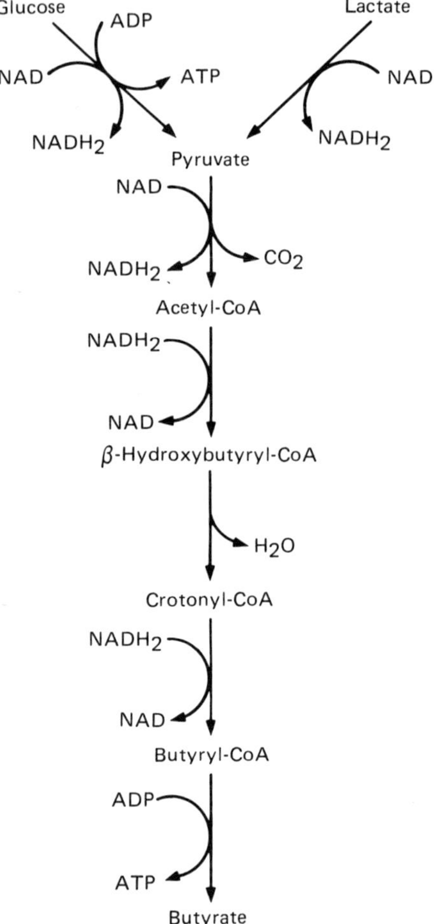


**Fig.4. Fermentation of glucose and lactate by the saccharolytic clostridial bacteria**

**Table 9. Products of the clostridial breakdown of amino acids in the silo†**

| Amino acid | Fermentation products |
|---|---|
| Aspartate | Fumarate, ammonia |
| Glutamate | Acetate, pyruvate, $\alpha$-aminobutyrate, $\gamma$-aminobutyrate, ammonia |
| Methionine | $\alpha$-Ketobutyrate, methyl mercaptan, ammonia |
| Serine | Pyruvate, ethanolamine, ammonia |
| Threonine | $\alpha$-Ketoglutarate, ammonia |
| Phenylalanine | $\beta$-Phenylethylamine, phenyl propionate, ammonia |
| Tyrosine | Tyramine, $\beta$-hydroxyphenyl propionate, ammonia |
| Tryptophan | Tryptamine, indole propionate, ammonia |
| Glutamine | Glutamate, ammonia |
| Asparagine | Aspartate, ammonia |
| Histidine | Histamine |
| Lysine | Acetate, butyrate, cadaverine, ammonia |
| Arginine | Putrescine |

† Reactions take place through: deamination (amino acid → organic acid + ammonia); decarboxylation (amino acid → amine + carbon dioxide); and exchange reactions (amino acid A + amino acid B → organic acid + ammonia + carbon dioxide)

acid-producing organisms increase dramatically (table 10) and the multiplication of organisms continues for the first few days in the silo, before the viable counts begin to decrease. The numbers of clostridial organisms also increase but their multiplication depends closely on the conditions of ensilage. Their growth is inhibited at high osmotic pressures (i.e. solute concentrations) and their sensitivity to this effect varies with the pH of the growth medium (fig. 5). The osmotic pressure of fresh grass is increased as the crop wilts and loses water, and the danger of a substantial clostridial growth recedes as the dry matter content of the crop approaches 300 g. kg$^{-1}$, the effect being enhanced progressively as the pH is reduced below 5.4. Yeasts are also present in most silages but they are normally quiescent and play little part in the ensilage process.

**Table 10. Microbial counts on fresh grass and silage samples (number of organisms per kg fresh material) (after McDonald, 1976)**

| | Total | Lactic acid bacteria |
|---|---|---|
| Uncut grass | $5.9 \times 10^9$ | $<10^5$ |
| Forage harvested grass | $2.5 \times 10^{11}$ | $4.9 \times 10^8$ |
| Grass at silo | $3.0 \times 10^{11}$ | $8.3 \times 10^7$ |
| Silage (after 189 d) | $3.9 \times 10^7$ | $7.0 \times 10^6$ |

However, they may become active after the silo has been opened, or if the seal on the silo is broken, and under these circumstances their action is often accompanied by fungal attack and spoilage (Beck, 1978).

Initially, when forage is placed in the silo, there is aerobic breakdown and oxidation of plant constituents through the action of both plant and microbial enzyme systems, and this leads to heating and a rise in temperature. As the crop is compacted and the silo sealed the oxygen supply is used up and anaerobic fermentation takes over. Examination of the crop during ensilage shows that there are major changes in composition. As might be expected, there is a rapid loss in the soluble carbohydrate fractions of the plant that form the main substrate for fermentation but there is also some degradation of hemicellulose (Dewar *et al.*, 1963; Morrison, 1979) and a marked redistribution of the plant nitrogenous fractions, reflecting hydrolysis of plant protein, and some degradation of amino acids (table 11). There is also a corresponding hydrolysis of plant lipids but with little modification of the fatty acids released (Jackson & Anderson, 1971; Lough & Anderson, 1973). Provided clostridial action is avoided, fermentation of soluble carbohydrate may be inhibited and a stable preserved silage obtained in a period of 2-3 weeks, although changes in composition, e.g. in the structural carbo-

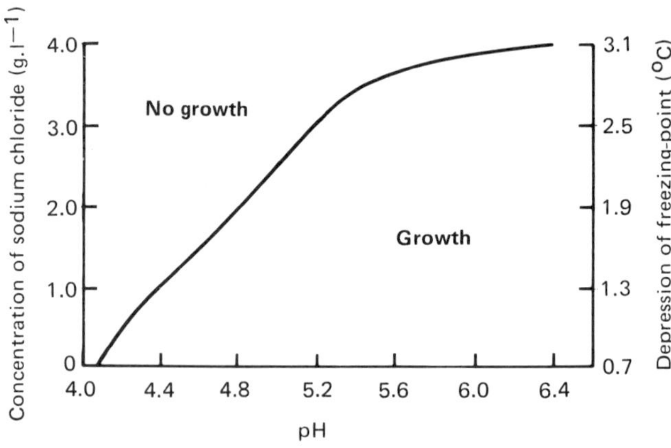


Fig.5. **The effect of osmotic pressure and pH on the growth of clostridial bacteria (after Wieringa, 1958).** Differences in osmotic pressure, measured by depressions of the freezing point, were created by varying the concentration of sodium chloride in the growth medium. Fresh grass has a depression of freezing-point of approximately 1.5°C and a natural pH of approximately 6.5

hydrate and protein fractions, may continue to take place after this period (see Carpintero *et al.*, 1979).

**Table 11. Nitrogen distribution in direct-cut and wilted ryegrass, and in silages made from the grass (after Brady, 1960)**

| | Nitrogen (g. $kg^{-1}$ total nitrogen) | | | | |
|---|---|---|---|---|---|
| | **Total soluble** | **Soluble non-protein nitrogen** | **Amino-nitrogen** | **Volatile nitrogen** | **Amide-nitrogen** |
| Fresh grass | 170 | 89 | 26 | 1 | 8 |
| Silage | 528 | 525 | 214 | 98 | 12 |
| Wilted grass | 189 | 173 | 92 | 6 | 13 |
| Silage | 556 | 548 | 274 | 69 | 5 |

## The efficiency of silage making

Losses in silage making occur in the field and in the silo during preservation and storage of the crop. The overall losses may be measured in chemical terms as the yield of constituent in grass minus the yield of constituent in silage, but there are technical difficulties in the determinations (see Watson & Nash, 1960; Nørgaard-Pedersen, 1975) and such measurements do not comment on the efficiency of the separate stages of silage making. This is important since it is clear that the losses vary substantially, not only between plant constituents but also with the silage making technique. Watson & Nash (1960), summarizing the literature, concluded that dry matter losses could be as high as 40% but were on average 12-19%. Corresponding figures for crude protein and for metabolizable energy were 11-20% and 20-30%, respectively. Recent studies (Papendick, 1974; Waldo, 1977; Lingvall, 1978) have shown a similar range of dry matter loss but they indicate that, with good ensilage techniques, dry matter losses may be restricted to values approaching 10%. These figures do not, of course, comment on the biological efficiency of the ensilage process, since they take no account of differences in nutritional value between grass and the silage.

Field losses in silage making arise through plant respiration, weather damage and inefficiencies in the harvesting of the crop. Respiration losses of dry matter will account, under many conditions, for approximately 1-2% of the plants' dry weight per 24 h, although the loss of nutritive value is somewhat greater since it is the soluble, highly digestible carbohydrate that is affected. Where weather conditions are poor, field losses rise sharply, particularly if wilting is prolonged (fig. 6). Rain leaching of nutrients from the crop is a special danger, which is exacerbated if there has been bruising or laceration to release the cell sap.

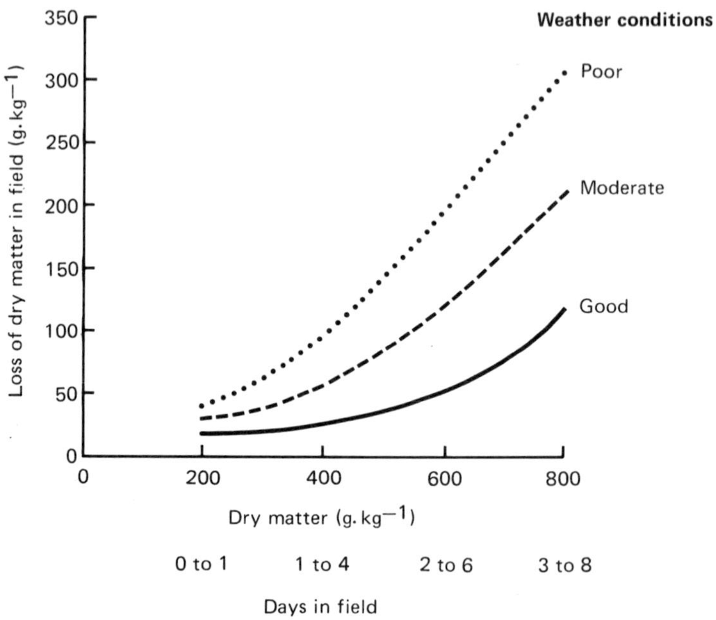


**Fig.6. Variations with weather conditions in field losses during conservation (adapted from Zimmer, 1977)**

Losses in the silo are due to plant and microbial aerobic metabolism, avoidable and unavoidable fermentation, and effluent. Prolonged aerobiosis following the initial ensilage of the crop results directly in inefficiency in conservation because of the oxidation of the plant sugars but there are other associated dangers. Aerobic conditions allow the development of clostridia and yeasts which lead to silage fermentations with poor preservation characteristics (table 12). Where air is satisfactorily excluded from the silo, anaerobic fermentation proceeds as indicated earlier but the efficiency of the process depends on the type of fermentation and the time taken to achieve the pH needed for preservation. Homolactic fermentations are the most efficient since they yield products of the maximum acidity with little loss of dry matter or energy. Heterolactic fermentations are less efficient, giving acid and neutral

**Table 12. The influence of oxygen in the silo on silage fermentation (from MacDonald, 1976)**

| | Grass | Silages | |
|---|---|---|---|
| Oxygen/grass in silo (l. $kg^{-1}$) | - | 0.26 | 6.59 |
| pH | 6.08 | 3.91 | 5.72 |
| Sugars (g. $kg^{-1}$ dry matter (DM)) | 124.9 | 9.9 | 12.5 |
| Volatile nitrogen (g. $kg^{-1}$ DM) | 0.3 | 2.7 | 9.6 |
| Acetate (g. $kg^{-1}$ DM) | - | 25.8 | 53.6 |
| Butyrate (g. $kg^{-1}$ DM) | - | 0 | 32.0 |
| Lactate (g. $kg^{-1}$ DM) | - | 14.42 | 0 |
| Total bacteria (x $10^{-9}$)† | 176 | 15.2 | 68.0 |
| Lactic acid bacteria (x $10^{-9}$)† | 0.72 | 25 | 132 |
| Yeasts and moulds (x $10^{-7}$)† | >10 | 0.7 | 31 |
| Clostridia | | | |
| Proteolytic | 250 | 1100 | 30 |
| Lactic fermenting | 30 | <10 | 250 |

† Microbial numbers per kg fresh weight

products, and a higher loss of dry matter, although energy loss differs little from that with homolactic fermentation because some of the fermentation products have a high energy content. Clostridial and yeast fermentations are especially inefficient (table 13).

The best conditions for ensilage, therefore, are those where preservation is achieved through a short period of homolactic fermentation. This is most likely where there is an adequate supply of soluble sugars to be fermented, and where there is an initial fall in pH, sufficiently large and rapid to discourage clostridial action. A general recommendation to obtain satisfactory fermentation is that forage should contain 60-80 g water-soluble carbohydrate per kg dry matter (McCullough, 1977) but this is an oversimplification. The rate of fall in pH in the silo depends not only on the production of acid but also on the buffering capacity of the crop itself, which is related to its contents of organic acid salts, phosphates etc., and varies with the type and species of forage, and with growth conditions and stage of growth (McDonald & Henderson, 1962). Moreover, since the effect of pH on clostridia depends on the dry matter content of the forage this should also be taken into account; the interrelationships involved have been summarized schematically by Weissbach and his colleagues (fig. 7).

An alternative approach to the problem is to ensile the forage with a chemical additive that preserves the crop directly by inhibiting microbial

**Table 13. Examples of the efficiency of some silage fermentations (after McDonald *et al.*, 1973)†**

| Type of fermentation | | Examples of dry matter (g) and energy (kJ) balances | | | | | | | Recovery of dry matter and energy in products‡ |
|---|---|---|---|---|---|---|---|---|---|
| Homolactic fermentation | | Glucose | + 2 ADP → | 2 lactate | + 2 ATP | | | | |
| | Dry matter | 180 | | 180 | | | | | 1.00 |
| | Gross energy | 2815 | | 2729 | | | | | 0.969 |
| Heterolactic fermentation | | Glucose | + ADP → | lactate | + ethanol | + $CO_2$ | + ATP | | |
| | Dry matter | 180 | | 90 | 46 | | | | 0.755 |
| | Gross energy | 2815 | | 1364 | 1371 | | | | 0.971 |
| | | 3 Fructose | + 2 ADP → | lactate | + acetate | + 2 mannitol | + $CO_2$ | + 2 ATP | |
| | Dry matter | 540 | | 90 | 60 | 364 | | | 0.951 |
| | Gross energy | 8478 | | 1364 | 876 | 6090 | | | 0.982 |
| Yeast fermentation | | Glucose | + 2 ADP → | 2 ethanol | + 2 $CO_2$ | + 2 ATP | | | |
| | Dry matter | 180 | | 92 | | | | | 0.511 |
| | Gross energy | 2815 | | 2742 | | | | | 0.974 |
| Clostridial fermentation | | 2 Lactate | + ADP → | butyric acid | + 2 $CO_2$ | + 2 $H_2$ | + ATP | | |
| | Dry matter | 180 | | 88 | | | | | 0.489 |
| | Gross energy | 2729 | | 2194 | | | | | 0.804 |

† Abbreviations: see fig. 3.
‡ Not including ATP energy (33.5 kJ. $mol^{-1}$)

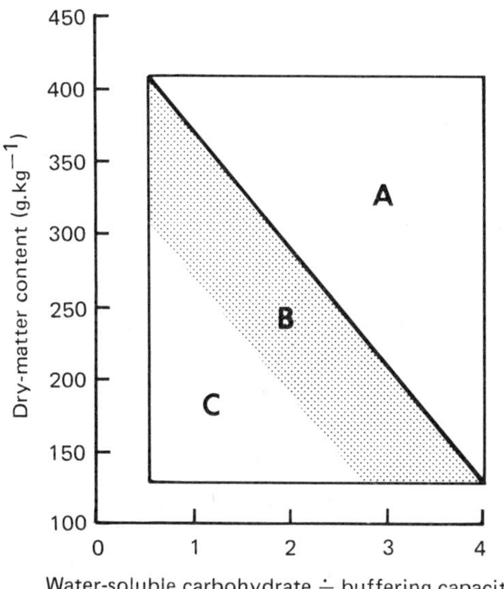


**Fig.7. A schematic representation of the interacting effects of crop dry matter, buffering capacity (g lactic acid per kg dry matter required to reduce pH to 4) and water-soluble carbohydrate contents (g. kg$^{-1}$ dry matter) on the quality of silage fermentation (after Weissbach *et al.*, 1974)** (A, good fermentation; B, uncertain fermentation; C, poor fermentation)

activity or that renders the conditions favourable for the development of the desired silage fermentation. The traditional AIV (A.I. Virtanen) mixtures, which contained strong mineral acids, were examples of additives of the inhibitor type, but almost all additives in use in Britain today are of the type designed to manipulate natural silage fermentation. With a typical modern additive, containing, for example, formic acid and/or sulphuric acid, sufficient is added to the crop to reduce the initial pH to approximately 4.6. This will not, in itself, give preservation but it does facilitate preservation through a short period of lactate fermentation. Some additives also contain formaldehyde, which reacts with grass protein, reducing its breakdown in the silo and, if the level of formaldehyde addition is high enough, in the rumen also (see Barry, 1976; Wilkinson *et al.*, 1976).

Effluent losses are a special problem with silages of low dry-matter content and may account for a substantial loss of soluble nutrients amounting to approximately 6-8% of the crop dry matter. The flow of effluent has been related to the dry-matter content of the crop by the equation:

$$W = 83.26 - 5.418D + 0.00883D^2$$

where $W$ is the weight of effluent as a percentage of the weight of the fresh herbage, and $D$ is the percentage of dry matter in the crop (see McDonald, 1976). In practice, effluent loss varies to a degree with the method of ensilage but with all methods the loss reduces to zero at a forage dry-matter content of between 25 and 30%.

In some circumstances there may be additional and substantial (up to 10%) losses in silage making through 'secondary oxidation' or 'aerobic deterioration' by bacteria, yeasts and fungi after the silo has been opened for use (Beck, 1978; Crawshaw & Llewelyn, 1978; Woolford, 1978). Where losses of this type occur they are predominantly with wilted silages containing more than 300 g dry matter per kg and a high level of residual water-soluble carbohydrate.

## REFERENCES

AGRICULTURAL RESEARCH COUNCIL 1976 *The Nutrient Requirements of Farm Livestock. No. 4, Composition of British Feedingstuffs.* Agricultural Research Council, London

BARRY, T.N. 1976 The effectiveness of formaldehyde treatment in protecting dietary protein from rumen microbial degradation. *Proceedings of the Nutrition Society* **35** 221-229

BECK, T. 1978 The micro-biology of silage fermentation. In *Fermentation of Silage — A Review* pp. 61-115 (Ed. M.E. McCullough). National Feed Ingredients Association, West Des Moines, Ia

BRADY, C.J. 1960 Redistribution of nitrogen in grass and leguminous fodder plants during wilting and ensilage. *Journal of the Science of Food and Agriculture* **11** 276-284

CARPINTERO, C.M., HENDERSON, A.R. & McDONALD, P. 1979 The effect of some pre-treatments on proteolysis during the ensiling of herbage. *Grass and Forage Science* **34** 311-315

CRAWSHAW, R. & LLEWELYN, R.H. 1978 The aerobic deterioration of grass silages. *Proceedings, 5th Silage Conference* (Ed. R.D. Harkess). pp 4-5 Hannah Research Institute, Ayr

DEWAR, W.A., McDONALD, P. & WHITTENBURY, R. 1963 The hydrolysis of grass hemicelluloses during ensilage. *Journal of the Science of Food and Agriculture* **14** 411-417

FRY, G. 1885 *Sweet Ensilage.* Agricultural Press Company, London

GOFFART, A. 1877 *Manuel de la Culture et de l'ensilage de Maïs et Autres Fourrages Verts* Masson, Paris

HAWKE, J.C. 1973 Lipids. In *Chemistry and Biochemistry of Herbage* (Eds G.W. Butler & R.W. Bailey) Vol. 1, pp 213-263 Academic Press, London

JACKSON, N. & ANDERSON, B.K. 1971 Effect of arachis oil on the conservation of heavily wilted herbage ensiled in plastic containers. II. — Fate of the arachis oil and its effect on the rumen volatile fatty acids of sheep. *Journal of the Science of Food and Agriculture* **22** 424-426

JOHNSTON, J.F.W. 1843 On the feeding qualities of the natural and artificial grasses in different states of dryness. *Transactions of the Highland and Agricultural Society of Scotland, 3rd Series* **1** 57-63

KEMBLE, A.R. & MACPHERSON, H.T. 1954 Liberation of amino acids in perennial rye grass during wilting. *Biochemical Journal* **58** 46-49

LINGVALL, P. 1978 Effects of different conservation systems on the utilization of lucerne. *Proceedings, 5th Silage Conference* (Ed. R.D. Harkess) pp 50-51 Hannah Research Institute, Ayr

LOUGH, A.K. & ANDERSON, L.J. 1973 Effect of ensilage on the lipids of pasture grasses. *Proceedings of the Nutrition Society* **32** 61A-62A (Abstr.)

LYTTLETON, J.W. 1973 Proteins and nucleic acids. In *Chemistry and Biochemistry of Herbage* (Eds G.W. Butler & R.W. Bailey) Vol. 1, pp 63-103 Academic Press, London

McCULLOUGH, M.E. 1977 Silage and silage fermentation. *Feedstuffs* **49**(13) 49-50

McDONALD, P. 1976 Trends in silage making. In *Microbiology in Agriculture, Fisheries and Food* (Eds F.A. Skinner & J.G. Carr) pp 109-123 Academic Press, London

McDONALD, P. & HENDERSON, A.R. 1962 Buffering capacity of herbage samples as a factor in ensilage. *Journal of the Science of Food and Agriculture* **13** 395-400

McDONALD, P., HENDERSON, A.R. & RALTON, I. 1973 Energy changes during ensilage. *Journal of the Science of Food and Agriculture* **24** 827-834

MORRISON, I.M. 1979 Changes in the cell wall components of laboratory silages and the effect of various additives on these changes. *Journal of Agricultural Science* **93** 581-586

NØRGAARD-PEDERSEN, E.J. 1975 [Determination of ensiling losses.] *Tidsskrift for Planteavl* **79** 561-608

PAPENDICK, K. 1974 Losses in the handling and conservation of forages. *Proceedings, 5th General Meeting of the European Grassland Federation, 1973,* Uppsala *Växtodling* **28** 81-89

SCHUKKING, S. 1976 The history of silage making. *Stikstof* **19** 2-11

SULLIVAN, J.T. 1973 Drying and storing herbage as hay. In *Chemistry and Biochemistry of Herbage* (Eds G.W. Butler & R.W. Bailey) Vol.3, pp 1-31 Academic Press, London

VIRTANEN, A.I. 1938 *Cattle Fodder and Human Nutrition with Special Reference to Biological Nitrogen Fixation.* Cambridge University Press, Cambridge

WAITE, R. 1965 The chemical composition of grasses in relation to agronomical practice. *Proceedings of the Nutrition Society* **24** 38-46

WAITE, R., JOHNSTON, MARGARET J. & ARMSTRONG, D.G. 1964 The evaluation of artificially dried grass as a source of energy for sheep. I. The effect of stage of maturity on the apparent digestibility of rye-grass, cocksfoot and timothy. *Journal of Agricultural Science* **62** 391-398

WALDO, D.R. 1977 Potential of chemical preservation and improvement of forages. *Journal of Dairy Science* **60** 306-326

WATSON, S.J. 1938 *Silage and Crop Preservation* Macmillan, London

WATSON, S.J. 1939 *The Science and Practice of Conservation: Grass and Forage Crops* 2 vols Fertiliser and Feeding Stuffs Journal, London

WATSON, S.J. & NASH, M.J. 1960 The losses involved in silages. In *The Conservation of Grass and Forage Crops* pp 319-395 Oliver and Boyd, Edinburgh

WEISSBACH, F., SCHMIDT, L. & HEIN, E. 1974 Method of anticipation for the run of fermentation in silage making, based on the chemical composition of green fodder. *Proceedings, 12th International Grassland Congress,* Moscow, pp 226-236

WIERINGA, G.W. 1958 The effect of wilting on butyric acid fermentation in silage. *Netherlands Journal of Agricultural Science* **6** 204-210

WILKINSON, J.M. 1981 Losses in the conservation and utilization of grass and forage crops. *Annals of Applied Biology* **98** 365-375

WILKINSON, J.M., WILSON, R.F. & BARRY, T.N. 1976 Factors affecting the nutritive value of silage. *Outlook on Agriculture* **9** 3-8

WOOLFORD, M.K. 1978 The aerobic deterioration of silage. *ARC Research Review* **4** 8-12

WYLAM, CLARE B. 1953 Analytical studies on the carbohydrates of grasses and clovers. III. — Carbohydrate breakdown during wilting and ensilage. *Journal of the Science of Food and Agriculture* **4** 527-531

ZIMMER, E. 1977 Factors influencing fodder conservation. *Proceedings, International Meeting on Animal Production from Temperate Grassland,* Dublin, pp 121-125

## Chapter 3

# The sward: its composition and management

**D. Reid**

*Hannah Research Institute, Ayr KA6 5HL*

A system for the production of herbage for making into silage has several aims. First, a satisfactory yield of digestible nutrients per ha must be obtained over the growing season, and this objective must be achieved taking account of the progressive increase in dry matter yield and reduction in digestibility that occurs as the crop matures. Secondly, a high proportion of the total yield should be produced early in the season when conditions are most suitable for the field operations involved. Thirdly, the herbage must have a chemical composition that allows a good pattern of fermentation in the silo to be achieved, so that the resulting silage is well preserved and has a high feeding value. To devise a herbage production system to meet these aims requires consideration of the effects of various factors on crop yield and composition. These factors, although often interacting, may be classified loosely under three headings: sward composition, cutting management and fertilizer application.

## SWARD COMPOSITION

In formulating a seeds mixture for a sward to be cut for silage, attention must first be paid to the species and variety of grass, and to whether a legume should be included.

### Choice of grass species

Judged on most criteria, perennial ryegrass (*Lolium perenne*) appears to be the ideal species to include in a conservation sward, although the published information shows some variation in its ranking among the persistent grasses in common use. McBratney & Laidlaw (1974) reported greater total yields from perennial ryegrass than from cocksfoot (*Dactylis glomerata*) and timothy (*Phleum pratense*) over a range of cutting frequencies and nitrogen application rates. However, the yields from tall fescue (*Festuca arundinacea*) were equal to or slightly greater than those from perennial ryegrass when cutting was infrequent, but smaller where the cutting interval was short. Corrall *et al.* (1979) confirmed both the superiority of tall fescue under infrequent cutting and the lower yields from timothy, but also showed slightly greater yields from cocksfoot than from ryegrass.

In contrast to the information on total dry matter yield, there is greater certainty that the average digestibility of perennial ryegrass is higher over the growing season than that of most other common grasses (Minson *et al.*, 1960; Green *et al.*, 1971). Perennial ryegrass should, therefore, generally give a greater annual yield of digestible nutrients than the other persistent grasses, but it would appear to differ little in this respect from the less persistent Italian ryegrass (*Lolium italicum*) (Corrall *et al.*, 1979).

For silage making the first cuts in the spring are of paramount importance, and here perennial ryegrass shows its superiority. Green *et al.* (1971) concluded that this grass outyielded other grasses at a given digestibility, and that on a given date it yielded 0.15 more dry matter than grasses of comparable digestibility. Although some varieties of cocksfoot reached a given digestibility at an earlier date, the early varieties of ryegrass gave higher yields at a high digestibility in early May. The maintenance of a high digestibility to a relatively late stage of development makes perennial ryegrass particularly valuable for silage making, since cutting may be continued over a longer period than with other grasses. Alternatively, cutting may be delayed until weather conditions are suitable for field operations.

A possible disadvantage of perennial ryegrass cut at a late stage is that it has a lower crude protein content than other grasses at the same digestibility level (Green *et al.*, 1971). Italian ryegrass has an even lower crude protein content than perennial ryegrass at a mature stage, but it is generally higher yielding at a given digestibility in the spring growth and decreases more slowly in digestibility as it matures. The higher moisture content of Italian ryegrass compared with perennial ryegrass is an additional disadvantage.

The content of soluble carbohydrate in the herbage is important in the establishment of the correct fermentation pattern in the silo and in this respect perennial ryegrass is again superior to other grasses. For example, Jones *et al.* (1961) found, in an early April first cut, that the soluble carbohydrate content of perennial ryegrass was 175 g. $kg^{-1}$, compared with 57 g. $kg^{-1}$ in cocksfoot where no nitrogen was applied. Similar differences were noted in mid-May and mid-June regrowths. Jones *et al.* (1961) also demonstrated that perennial ryegrass had a higher soluble carbohydrate content than cocksfoot at each successive cut through the season until late September.

### *Selection of ryegrass variety*

Accepting perennial ryegrass as the ideal species for silage swards, the relative merits of the various varieties of this grass should be considered. Results presented by Corrall *et al.* (1979) show that the early variety, S24, slightly but consistently exceeded the late variety, S23, in total dry matter yield over the season. However, S23 had a higher average digestibility and crude protein content. In contrast, Lee *et al.* (1977) showed that S23 gave a slightly greater total yield of dry matter than S24. This experiment included several varieties of ryegrass, and all the late varieties gave greater total yields and higher crude protein contents than the early ones. The very late variety Melle was the highest yielding, but it showed no advantage over the late S23 or the intermediate Barlenna in crude protein content.

The most extensive information on differences between varieties of ryegrass at the primary growth in the spring is provided by Green *et al.* (1971) and Corrall *et al.* (1979). On average, at a given date in the spring, the late variety, S23, gave a lower yield of herbage but with a higher digestibility than the early variety, S24. However, S23 had a lower digestibility at the same stage of development. The tetraploid varieties, Reveille and Petra, which have similar maturity dates to S24 and S23 respectively, gave slightly lower yields but higher digestibilities at a given date in the spring than their diploid counterparts. After mid May the tetraploid and diploid varieties gave almost the same yield of digestible organic matter per unit area. One advantage of the tetraploid varieties is their high content of soluble carbohydrate. Thus in herbage cut for silage in May and August, Castle & Watson (1971) reported carbohydrate contents of 137 and 182 g. kg$^{-1}$ respectively for Reveille, compared with 98 and 131 g. kg$^{-1}$ respectively for the diploid S24. However, the disadvantage of high moisture contents in the tetraploid varieties has also been demonstrated (Green *et al.*, 1971).

## The role of legumes

The next point to be discussed with regard to seeds mixture for a silage sward is whether or not to include a legume with the grass, or even to sow a legume alone. Aldrich (1974) drew attention to the marked shift in Britain, over the 10 years prior to 1974, from the growing of legumes towards the use of fertilizer nitrogen. He argued for a reversal of this trend in view of increasing fertilizer prices and the introduction of improved varieties of legumes, and he also suggested that much more use should be made of lucerne and red clover as conservation crops. The main factor making this suggestion possible was the introduction of disease-resistant varieties of these legumes. In addition, the poor ensiling characteristics of these crops could now be overcome by the use of silage additives. However, in many grassland areas lucerne is still an uncertain crop because of its intolerance to acid soil conditions and high water tables. Red clover appears to be a more successful crop and has received renewed attention in the last few years.

Dickinson (1972) reported that tetraploid varieties of red clover were more persistent than diploid varieties. The yield from tetraploids was slightly greater than that from diploids in the 1st harvest year and remained at about the same level into the 2nd year, when the yield from the diploids decreased markedly. By the 3rd year only the tetraploids were productive, although giving a lower yield than in the previous year. Hunt *et al.* (1975) confirmed the high productivity of tetraploid varieties of red clover, but found that tetraploidy did not guarantee persistency. Some varieties were productive for 1 harvest year, others for 2, but few maintained productivity for 3 years. The most productive tetraploid variety in the trials of Hunt *et al.* (1975) was Hungaropoly. In small-plot experiments, swards of this variety gave total dry matter yields of 11.7 and 14.1 t. $ha^{-1}$ in the 1st and 2nd harvest years respectively, with red clover contents of 0.80 and 0.99. However, the yield dropped sharply to 8.8 t. $ha^{-1}$ with 0.82 clover in the 3rd year.

When used as a silage crop on a field scale, Hungaropoly red clover appears to be less persistent than the above results from plot experiments indicate. Thus, one field of this clover (Castle & Watson, 1974) gave a total dry matter yield of 13 t. $ha^{-1}$ in the 1st harvest year, but only 8.7 t. $ha^{-1}$ in the 2nd; so few plants of red clover remained in the autumn of the 2nd year that the field was ploughed. This lack of persistency in the field was attributed to mechanical damage by the wheels of tractors and other silage-making equipment. Frame (1975) suggested that wheel damage might be reduced by sowing a companion grass and recommended a medium-late to late perennial ryegrass at a low seed rate. This might also improve the nutritive value of the herbage and reduce weed infestation.

Because of the lack of persistency and the consequent need for resowing every 2-3 years, the place of red clover in a conservation system remains doubtful and the apparently superior variety, Hungaropoly, is not now recommended because of its high susceptibility to red clover mosaic virus. In the past white clover has been considered suitable only under grazing conditions, but recent information suggests that the new large-leaved varieties might be of value under cutting conditions also. Because of its growth habit white clover should be more resistant to wheel damage in the field than red clover. In a recent plot experiment (D. Reid, unpublished results) the large-leaved variety of white clover, Blanca, grown alone gave total dry matter yields of 10.8, 9.3 and 7.2 t. $ha^{-1}$ in the 1st, 2nd and 3rd harvest years. These yields are lower than those observed in experiments with the best varieties of tetraploid red clover, but there was still a satisfactory population of white clover plants in the autumn of the 3rd year. It is certain, therefore, that productivity would have been maintained in later years without reseeding, which would have been necessary with red clover.

A further advantage of white clover is its high digestibility. Harkess (1963) showed that *in vivo* determinations of the digestibility of herbage from a pure sward of S100 white clover were consistently higher than from perennial ryegrass and cocksfoot. Further, the digestibility of the clover decreased with maturity by only 1.5 compared with 5 g. $kg^{-1}$ . $d^{-1}$ with ryegrass. According to Thomson & Raymond (1970), white clover is unique in these two features and these authors attributed this to the mode of growth in the white clover sward. A succession of petioles emerging above the existing canopy is accompanied by rapid senescence of shaded leaves, resulting in immature herbage being almost continuously available for cutting. In contrast, the erect-growing red clover has a lower digestibility than most grasses (Castle & Watson, 1974) and this is related to higher fibre content.

Thomson & Raymond (1970) stressed that, although the mode of growth of white clover is advantageous in relation to digestibility, it also infers a more rapid attainment of maximum yield per unit area, which will occur when the production of new tissue is equalled by the senescence of old tissue. In pure seedings this fact might be a disadvantage to the use of white clover, since more frequent cutting would be necessary than with red clover. The role of white clover in the production of herbage for silage making is most likely to be in mixtures with grasses. The main advantages of grass plus clover swards, particularly where the clover is a large-leaved variety, are the considerable reduction in nitrogen fertilizer rate required to achieve a given yield level (Reid, 1977) and the increase in the digestibility of the herbage (Harkess, 1963). The main disadvantage is that white clover has a shorter growing season than most grasses, being later to start active growth in the spring and earlier to stop in the autumn. However, this is likely to be less important in silage making than in grazing systems, since weather conditions in the spring and in the autumn are often unsuitable for cutting and wilting operations.

## CUTTING MANAGEMENT

The factors included in this category are: date of cutting the primary growth of the season, frequency of cutting, closeness of cutting and the effects of winter grazing.

### Date of cutting primary growth

The growth rate of perennial grasses is at a maximum during the period of stem elongation in the spring and early summer, and the herbage yield at this time contributes a large proportion of the total production for the season. Weather conditions in the spring and early summer are most suitable for silage making, so it is important to obtain maximum production and quality at this time. In general, the yield of herbage increases and its digestibility decreases as the date of cutting the primary growth is delayed. The exact rates of change vary with the species and variety of grass. These variations have been summarized in an earlier section, where the superior characteristics of perennial rye-grass compared with other grasses, relating to yield and digestibility in the primary growth, were demonstrated (Green *et al.,* 1971).

Typical effects of increasing the lateness of cutting on the primary growth from two contrasting varieties of perennial ryegrass are shown in fig. 1. The data are derived from a trial of Corrall *et al.* (1979), in which the primary growth from the late variety S23 and the early variety S24 was cut at one of six dates between mid April and late June. Herbage from S23 had a higher digestibility than that from S24 at all except the earliest cutting dates. In addition, the digestibility of S23 declined less rapidly than that of S24 up to late May, but more rapidly thereafter. The decline of S24 was almost linear at an average rate of 3.3 g. $kg^{-1}.d^{-1}$ over the whole period, but that with S23 averaged 1.6 g. $kg^{-1}.d^{-1}$ up to late May and 4.2 g. $kg^{-1}.\ d^{-1}$ over the next 4 weeks. Dry matter yield increased as the cutting date was delayed, S24 giving the higher yields when first cut in April or May, but almost the same yield as S23 when cut in June. The yields of digestible organic matter from S24 were also higher than those from S23 except at the June cuts, when they were equal. However, the increase in these yields from S24 was very slow

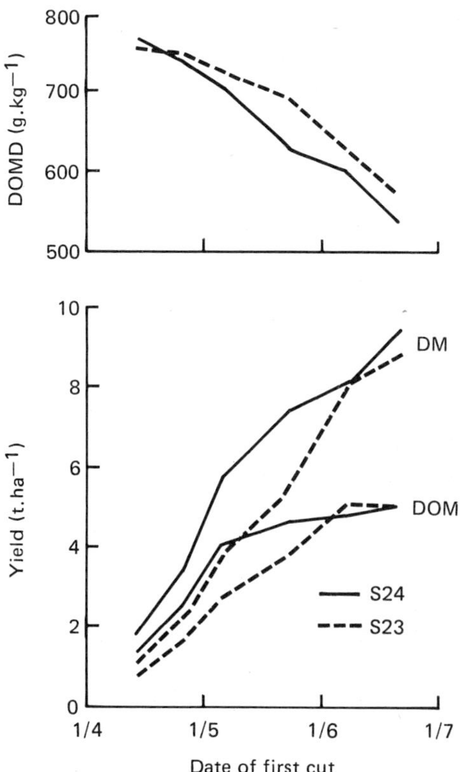


**Fig.1. Effects of date of cutting on yield and digestibility of primary growth from S23 and S24 perennial ryegrass (after Corrall *et al.*, 1979)** (DOMD, digestible organic matter in the dry matter; DM, dry matter; DOM, digestible organic matter)

after early May, whereas the increase in those from S23 was almost linear up to early June. These results confirm those from earlier trials (Green *et al.*, 1971). S23 perennial ryegrass will therefore provide herbage of a high digestibility over a longer period than will S24 ryegrass, the digestible organic matter content of the dry matter (DOMD) of the former remaining above 700 g. $kg^{-1}$ until approximately 10 days after the latter. However, the yield from S23 is lower at these high digestibility levels. Estimates suggest that S23 requires a further 7 days of growth to give the same yield as given by S24 at a DOMD content of 700 g. $kg^{-1}$ and by then the digestibility of S23 has decreased to 670 g. $kg^{-1}$.

The date of cutting the primary growth affects not only the yield and digestibility of that growth but also those of the secondary growth. Generally, yield decreases and digestibility increases slightly at the second cut as the date of the first cut is delayed. For example, in 1 year of the experiments of Green *et al.* (1971) the dry matter yields in regrowths from S23 ryegrass 50 days after first cutting on 19 April and 6 June were 7.1 and 2.6 t. $ha^{-1}$ respectively, with DOMD contents of 670 and 740 g. $kg^{-1}$. Comparable results for S24 ryegrass were 6.5 and 3.1 t. $ha^{-1}$ with DOMD contents of 630 and 700 g. $kg^{-1}$.

At the earliest date of cutting the primary growth in experiments of this type the developing flowering shoots on the ryegrass are immature and too short to be removed by mowing. Thus the secondary growth contains a high proportion of stem and emerged ears, resulting in a high yield of relatively low digestibility. In contrast, at the latest date of cutting the primary growth the majority of fertile shoots have emerged ears, all of which are removed by the mower. As a result, the secondary growth is mainly vegetative, with a low yield and a high digestibility. Between these two extremes the stage of development of flowering shoots and the proportion of them removed in the first cut appear to have interacting effects in the secondary growth. This leads to some variability in the yield and digestibility effects at the second cut, as shown by Corrall *et al.* (1979) for both S23 and S24 ryegrass.

On average, the effects of the date of first cutting dominate seasonal production, so that the total dry matter yield increases and weighted mean digestibility decreases with delaying the first cut (e.g. Green *et al.*, 1971). However, inconsistencies occur, which are again attributable to the interaction effects of stage of development and proportion of emerged ears removed at the first cut.

## Frequency of cutting

Averaged over 3 years and a wide range of nitrogen rates, McBratney & Laidlaw (1974) found that the total dry matter yields from four grasses increased as the interval between cuts was increased from 3 to 6 weeks. For example, the yields from S24 ryegrass at these two cutting intervals were 11.30 and 14.13 t. $ha^{-1}$ respectively. The number of cuts was decreased from 11 to 6 as the cutting interval increased from 3 to 6 weeks. The mean increase in yield with each decrease in the number of cuts was therefore 0.57 t. $ha^{-1}$.

Chestnutt *et al.* (1977) compared six cutting frequencies on S24 ryegrass swards for 3 years at two nitrogen rates and on three sites. Cutting interval varied from 2 to 8 weeks, giving between 15 and 4 cuts per year, and the mean annual yields of dry matter increased from 8.99 to 14.24 t. $ha^{-1}$. The yield increase with each decrease in the number of cuts thus averaged 0.48 t. $ha^{-1}$, slightly smaller than the previous estimate. The dry-matter digestibility decreased from 790 g. $kg^{-1}$ with cutting intervals of 2 weeks to 670 g. $kg^{-1}$ with intervals of 8 weeks. There was also a decrease in digestibility of approximately 0.3 g. $kg^{-1}$. $d^{-1}$ between successive cuts over the season and this varied little with the cutting frequency. The conclusion was that little benefit could be gained in terms of digestible organic matter yield per ha from increasing the interval between cuts to more than 4 weeks.

In the above experiments cutting frequency was determined on the basis of fixed time-intervals. However, Collins & McCarrick (1969) suggested that this method is applicable only in grazing studies and that the more desirable basis to use in conservation investigations is a predetermined stage of growth or height of herbage. Thus in their experiment the herbage was cut at heights varying from 50 to 300 mm, giving between 13 and 2 cuts each season, and the yield increased on average by 0.4 t. $ha^{-1}$ with each decrease in cut number.

Reid (1978) also adopted this system, cutting an S23 ryegrass sward at heights of 80-100, 150-200 or 250-300 mm, giving 10, 5 or 3 cuts respectively each year. Averaged over 3 years and a wide range of nitrogen rates, the total yields of herbage dry matter were 6.32, 10.03 and 11.25 t. $ha^{-1}$ with 10, 5 and 3 cuts respectively. Thus the mean yield increase was 0.70 t. $ha^{-1}$ for each decrease in cut number, a value which exceeds the estimates from previous experiments. Some variation in these estimates would be expected since they have been calculated as averages over different ranges of nitrogen rate, and interaction effects between cutting interval and nitrogen rate have been demonstrated (see page 52). Fig. 2 illustrates the average effects in the four references cited above.

The limited digestibility information reported by Reid (1978) from the above experiment agrees well with that of Chestnutt *et al.* (1977) in showing a decrease in digestibility with increasing cutting interval. The organic matter digestibility decreased from 760 g. $kg^{-1}$ with 10 cuts to 750 with 5 and 720 g. $kg^{-1}$ with 3. However, the seasonal trends in digestibility varied with cutting interval in this experiment but not in that of Chestnutt *et al.* (1977). With 10 cuts, herbage digestibility varied in an apparently random manner by approximately 10 g. $kg^{-1}$ about the mean. In contrast, with five and three cuts, digestibility decreased from cut to

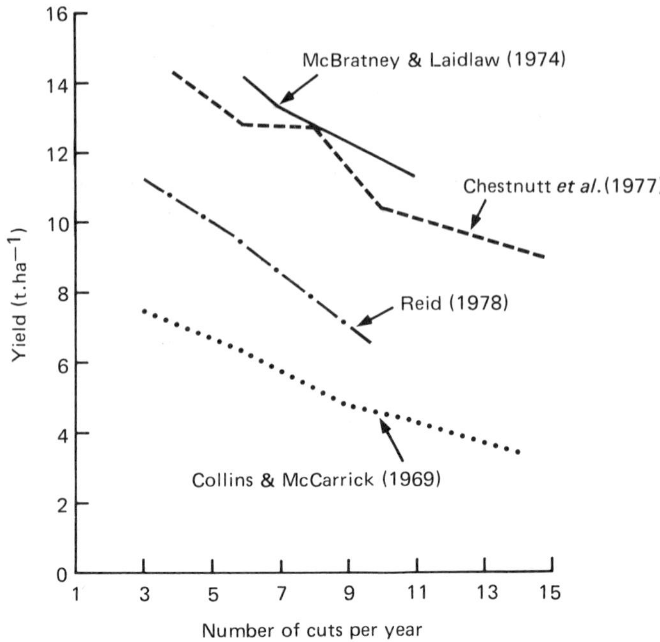


**Fig.2.** **Effects of frequency of cutting on total yield of herbage dry matter**

cut over the season at mean rates of 0.5 and 1.3 g. $kg^{-1}$ respectively. The benefit in terms of total yield of digestible organic matter obtained by decreasing the number of cuts to less than five was slight. This is equivalent to an average cutting interval of approximately 4 weeks, which is in agreement with the conclusions of Chestnutt *et al.* (1977).

In most experiments on frequency of cutting the effects of this factor were confounded with the effects of date of cutting the primary growth, this date being increasingly delayed as the interval between succeeding cuts was increased. For example, the mean dates of first cuts in the experiment of Reid (1978) were 2 May, 18 May and 1 June for the 10-, 5- and 3-cuts treatments respectively. Preliminary results from a current experiment (D. Reid, unpublished results) suggest that the effects of these two factors interact. In this experiment the primary growth on an S23 sward was cut at one of six dates 1 week apart between 14 May and 18 June 1979. The succeeding regrowths from each of these first cuts were then cut at intervals of 4 or 6 weeks until the end of the season. Fig. 3 shows that, although the total yield of dry matter increased as the date of first cut was delayed with both cutting intervals, the average increase was greater where the interval was 4 rather than 6 weeks. As a result, the yield difference in favour of cutting at 6-week intervals following early first-cut dates decreased, and there was little difference between the two frequencies following first cuts on 11 and 18 June. Interaction effects were also noted in the digestibility results. When the regrowths were cut every 4 weeks the mean DOMD decreased rather steadily from 780 to 730 g. $kg^{-1}$ as the date of first cut was delayed from 14 May to 18 June. However, with cutting intervals of 6 weeks the DOMD first increased from 700 to 760 g. $kg^{-1}$ as the first cut was delayed from 14 to 28 May and thereafter decreased in a similar fashion to that described for the other

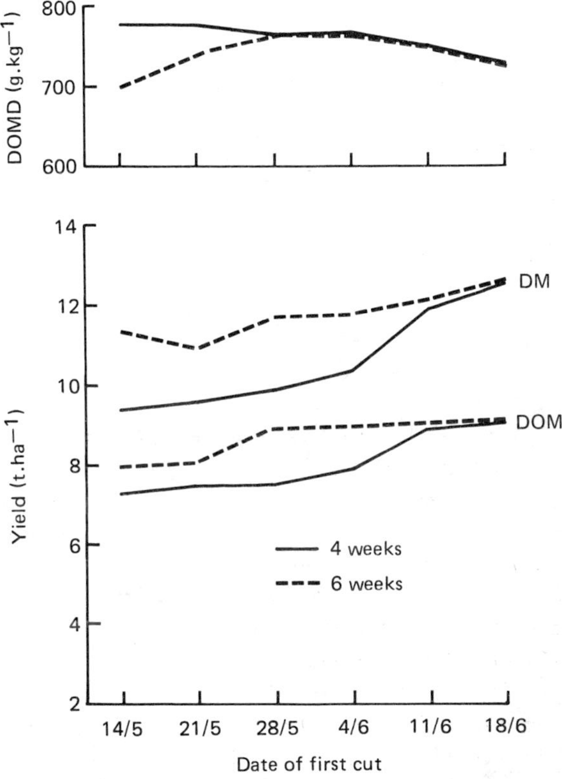


**Fig.3. Effects of date of first cut and subsequent cutting interval on total yield and average digestibility of herbage from S23 ryegrass**

cutting interval. In terms of yield of digestible organic matter there was, therefore, no benefit with a 6-week frequency in delaying the first cut beyond 28 May, whereas delaying until 11 June was advantageous with the 4-week frequency. The total yield of digestible organic matter did not, in fact, vary with the cutting interval where the primary growth was cut on 11 and 18 June. Similar interaction effects were noted in the results from this experiment in 1980 despite a considerable difference in the seasonal rainfall pattern.

## Closeness of cutting

Cutting closer to the ground level has been shown consistently to increase the total yields of dry matter from perennial ryegrass swards. For example, Reid (1959) obtained increases of 0.39-0.49 by cutting to 25 mm rather than to 60 mm. The increases were slightly smaller in a later experiment (Reid, 1966) in which the frequency of cutting was shown to have no influence on the effects of closeness of cutting. Another experiment (Reid, 1962) investigated the effects of continuing the cutting treatments on an S23 ryegrass sward over a period of 5 years and showed no reduction in the benefits of close cutting with time. In fact, the yield advantage of close cutting was greatest in the 4th and 5th years. The effects on herbage digestibility were not investigated in this study but Binnie & Harrington (1972) showed that the digestibility of Italian ryegrass was not affected by varying the closeness of cutting, although the dry-matter yield response of this grass was the same as noted for perennial ryegrass.

The practical application of these results is difficult since the experimental plots were cut with reciprocating-blade mowers. This type of mower is rarely used in the field nowadays, as most mowers are of the drum-disc or flail types. Little information is available on the effects of different types of mowers on herbage production. That the type of mower may have an effect is indicated by the results of Reid & MacLusky (1960), which showed that the yields from a perennial ryegrass sward cut to 25 mm were 0.035 to 0.125 greater when cut with a cylinder mower than with a reciprocating-blade mower.

Results reported by Black & Alexander (1967) suggested that the regrowth rates following close cutting were slower where a flail mower rather than a reciprocating-blade mower was used. However, this was not confirmed in a more recent experiment (D. Reid, unpublished results) in which a perennial ryegrass sward was cut to 25 mm with either a flail mower or a reciprocating-blade mower. No differences in growth rate were noted in four of the five growing periods in the season but, following an early June cut, the rates were greater with the flail mower. In an accompanying experiment a ryegrass sward gave a 0.20 greater yield of dry matter over the season when cut with a flail mower than with a reciprocating-blade mower. With both machines the yields were greater when cutting was to 25 mm rather than 60 mm, but the difference was only 0.10 with the flail mower compared with 0.16 with the reciprocating-blade mower. Observations suggest that the performance of the drum-disc mower is similar to the flail mower.

The conclusion from plot experiments is that closer cutting will increase herbage production, but there may be difficulties in adopting close cutting in practice. Even with careful cultivation methods the soil surface in a grass field will have some irregularities and, if cutting is too close, the sward will be scalped in places, damaging both sward and mower. The cut herbage will also be contaminated with soil, which will affect the fermentation pattern in the silo and the acceptability of the silage to the animal. In establishing a sward for conservation every effort should be made to prepare a level and stone-free seed bed, and to maintain this by periodic harrowing and rolling of the sward.

### Winter grazing

Two advantages of grazing a conservation sward during the winter months are the possible reduction in winter kill of the grasses and the removal of dead herbage, which might otherwise lower the digestibility of the spring crop. A disadvantage is the slower growth rate of the sward in the spring, particularly after late-winter grazing. However, the available evidence (e.g. Frame, 1965) suggests that this effect is most evident in swards used for early grazing and that, by normal silage-cutting dates, the yields are only slightly less than if the sward had not been grazed in the winter. This reduction in yield may be corrected readily by a small increase in the spring dressing of nitrogen fertilizer.

## FERTILIZER APPLICATION

The most important fertilizer nutrient controlling the production of herbage is nitrogen. This section concentrates, therefore, on the effects of nitrogen application on herbage yield and quality, and discusses the principal factors influencing these effects.

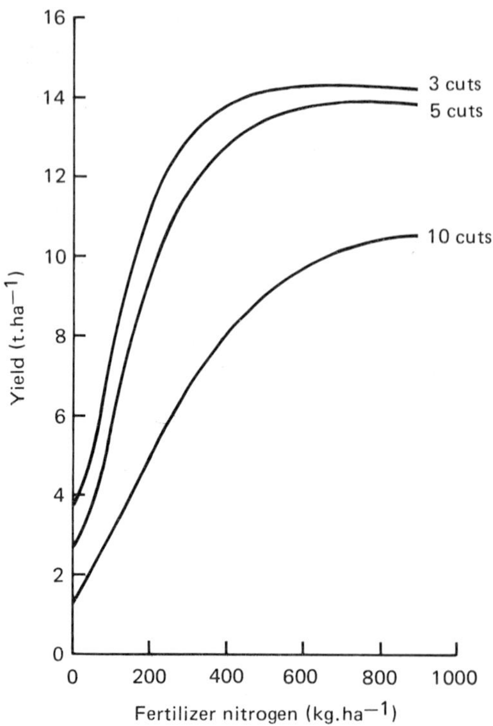


**Fig.4. Effects of frequency of cutting and nitrogen application rate on total yields of herbage dry matter from S23 perennial ryegrass (Reid, 1978)**

## Nitrogen rate and cutting frequency

Many workers (e.g. Chestnutt *et al.*, 1977; Reid, 1978) have shown that the frequency of cutting has a considerable influence on the response of total herbage yields to nitrogen application rate. Typical effects are illustrated in fig. 4, showing the results for an S23 perennial ryegrass sward in a 3-year experiment (Reid, 1978). When the sward was cut five times annually, at an average height of 150-200 mm, the mean annual yields of dry matter increased almost linearly as the nitrogen rate was increased from 0 to 340 kg. $ha^{-1}$ , each kg of nitrogen giving an extra 30 kg of dry matter. The response then declined to insignificant values with rates at and above 560 kg nitrogen per ha. In contrast, with three cuts annually, at a height of 250-300 mm, the response was linear only up to

280 kg nitrogen per ha, and then decreased more rapidly to reach insignificant values at and above 450 kg. $ha^{-1}$ , but the response to the lower rates was slightly greater at 34 kg dry matter per kg nitrogen applied. Less frequent cutting gave greater yields, on average, but the differences were small with nitrogen rates above 340 kg. $ha^{-1}$. Cutting 10 times annually at a height of 80-100 mm gave lower total yields at all nitrogen rates, but the response to applied nitrogen remained almost linear at 15 kg dry matter per kg nitrogen up to a nitrogen rate of 500 kg. $ha^{-1}$. The subsequent decline in response here was much less rapid than with five or three cuts.

Mathematical models were constructed from the results in each harvest year of this experiment, relating yield to nitrogen rate and cutting frequency. Predictions are given in table 1 of the amounts of nitrogen required in the 1st harvest year to produce given total yields of dry matter at several cutting frequencies. As the frequency of cutting is increased the nitrogen rate required to produce a given total yield also increases. In addition, the extra nitrogen required to produce a given increment of yield increases. For example, to increase yield from 9 to 13 t. $ha^{-1}$ would need an extra 90 kg nitrogen per ha with three cuts but 141 kg. $ha^{-1}$ with six cuts.

The proportional organic matter digestibility of the herbage in this experiment did not vary with nitrogen rate, but increased as cutting frequency increased. As a result, the difference in total yield of digestible organic matter between frequencies of three and five cuts was slight at all nitrogen rates.

A common criticism of systems designed to produce high-digestibility herbage is that they inevitably produce lower total yields of dry matter per unit area than where a lower digestibility herbage is accepted. However, the predictions in table 1 indicate clearly that this apparent problem may be solved by increasing the nitrogen application rate. In the main silage-making period, up to approximately mid July, estimates suggest that an increase of less than 0.10 in the nitrogen rate would

**Table 1. Nitrogen application rate (kg. $ha^{-1}$) required to produce given total yields of herbage dry matter (t. $ha^{-1}$) at four different cutting frequencies**

| Target dry matter yield (t. $ha^{-1}$) | No. of cuts per year | | | |
|---|---|---|---|---|
| | 3 | 4 | 5 | 6 |
| 9 | 80 | 95 | 112 | 134 |
| 11 | 121 | 141 | 165 | 195 |
| 13 | 170 | 197 | 231 | 275 |

compensate for the lower yield from cutting more frequently to achieve high digestibility. At the same time, a higher yield of digestible organic matter would be obtained per unit area.

In general, the application of nitrogen depresses the water-soluble carbohydrate content of grasses (Waite, 1958; Jones *et al.*, 1961) and this might be expected to affect the fermentation characteristics in the silage. However, perennial ryegrass has an inherently higher carbohydrate content than most other grasses, so the application of nitrogen has less serious consequences. Thus, Jones *et al.* (1961) showed that S23 ryegrass had, on average, a slightly higher carbohydrate content at a nitrogen rate of 90 kg. $ha^{-1}$ for each crop than had S143 cocksfoot with no applied nitrogen. Results reported by Reid & Strachan (1974) suggest that the relationship between carbohydrate content and nitrogen applications varies with the actual rate applied. Averaged over 3 years, the carbohydrate content of herbage from S23 ryegrass swards cut five times annually increased from 200 g. $kg^{-1}$ with no applied nitrogen to 220 g. $kg^{-1}$ with 112 g nitrogen per ha and then decreased slightly to 210 g. $kg^{-1}$ with 224 kg. $ha^{-1}$. With higher nitrogen rates the decrease was steady, reaching 200 g. $kg^{-1}$ at 250 kg. $ha^{-1}$ and 110 g. $kg^{-1}$ at 896 kg. $ha^{-1}$.

Jones *et al.* (1961) noted that the decrease in the carbohydrate content of herbage with the application of nitrogen was accompanied by an increase in the nitrogen content. Reid & Strachan (1974) investigated this relationship in S23 ryegrass swards with and without S100 white clover and showed that the water-soluble carbohydrate content decreased by 1 g. $kg^{-1}$ for each g. $kg^{-1}$ increase in the crude protein content. The presence or absence of white clover in the sward had no effect on the relationship, suggesting that it was attributable to the total nitrogen supply and not to the nitrogen rate as such. The clover and soil contributions to nitrogen supply must be taken into account, therefore, when assessing the carbohydrate status of herbage for silage making.

There is evidence to suggest that these carbohydrate effects have no influence on fermentation pattern in the silo, at least where the herbage is from a perennial ryegrass sward. Jones (1970) applied 50,100 or 200 kg nitrogen per ha on an S24 ryegrass sward and cut the primary growth either 1 week before or 3 weeks after 0.50 ear emergence. The carbohydrate contents of the herbage from the three nitrogen rates were 150, 130 and 100 g. $kg^{-1}$ respectively at the early date, and 150, 120 and 80 g. $kg^{-1}$ at the late date. These variations with nitrogen rate had no effect on fermentation pattern and all the silages were well preserved.

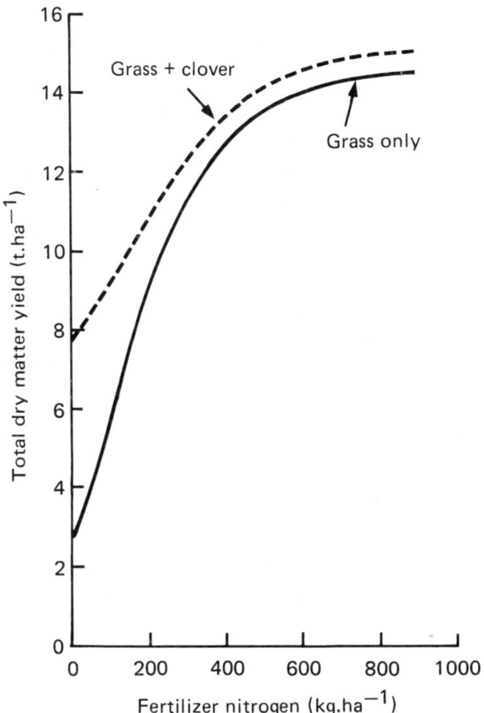


**Fig.5. Effects of fertilizer nitrogen rate on total yields of dry matter from S23 ryegrass, with and without S100 white clover**

## Nitrogen rate and white clover

The presence or absence of white clover in the sward has a marked influence on the response of herbage yields to fertilizer nitrogen rate, at least when the herbage is cut at a relatively immature stage (Reid, 1970 & 1972). In general, the addition of white clover increases total herbage yield and reduces the yield response to nitrogen at lower application rates, but has little effect at high rates. Fig. 5 shows these typical effects of the inclusion of S100 white clover in an S23 ryegrass sward over a nitrogen range from 0 to 896 kg. ha$^{-1}$ year$^{-1}$, and cut five times per year (Reid, 1970). Averaged over the first 3 harvest years the total dry matter yields were 7.9 and 2.8 t. ha$^{-1}$ from the grass plus clover and grass-only swards respectively where no nitrogen was applied. Although the yield response was almost linear between 0 and 340 kg nitrogen per ha on both swards, the increase per kg of nitrogen applied was only 14 kg of dry matter on the grass plus clover sward compared with 27 kg on the grass-only sward. Thus the yield difference between the two swards was slight with nitrogen rates at and above 340 kg. $ha^{-1}$. With no applied nitrogen the grass plus clover sward gave the same yield as the grass-only sward, with 190 kg nitrogen per ha.

Replacing S100 with a larger-leaved and taller-growing variety of white clover reduces the competition for light between the clover and the grass, resulting in a greater clover contribution to yield. Yield response curves for S23 ryegrass swards with and without the large-leaved variety Blanca are shown in fig. 6 (D. Reid, unpublished results). Where no nitrogen was applied the total dry-matter yields were 10.9 and 5.4 t. $ha^{-1}$ with and without this clover. Although the response rates in the linear sections of the curves in fig. 6 were both lower than those in fig. 5, the differences between them were similar. The increase per kg of nitrogen

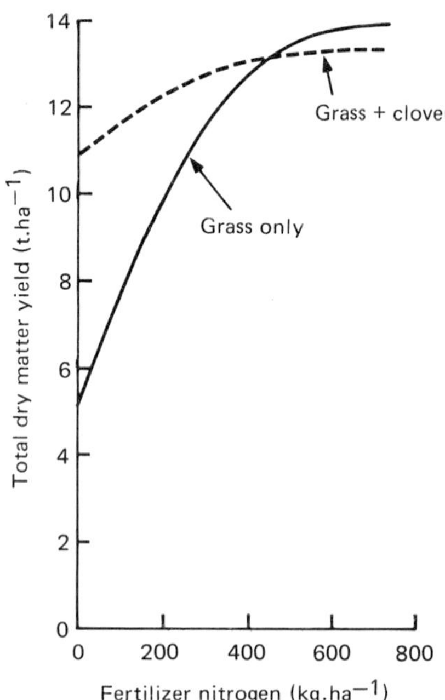


**Fig.6. Effects of fertilizer nitrogen rate on total yields of dry matter from S23 ryegrass, with and without Blanca white clover**

applied was only 6 kg of dry matter on the grass plus clover sward compared with 20 kg on the grass-only sward. With no applied nitrogen the grass plus Blanca sward gave the same yield as the grass-only sward, with 260 kg nitrogen per ha.

Under a less frequent cutting schedule the effects of including white clover might be reduced by the greater competition from the grass, particularly if the clover is a medium-leaved variety such as S100. However, with more frequent cutting for high digestibility herbage the inclusion of white clover should allow economy in nitrogen fertilizer input. Predictions suggest that to produce an annual yield of 12-13 t dry matter per ha would require a total nitrogen application of 360 kg. $ha^{-1}$ on a ryegrass-only sward, 260 kg. $ha^{-1}$ on a ryegrass plus S100 sward and only 180 kg. $ha^{-1}$ on a ryegrass plus Blanca sward. The potential savings in nitrogen fertilizer are therefore 0.20 with S100 and 0.50 with Blanca.

Another possible benefit from the inclusion of white clover is an increase in herbage digestibility. However, information on this is limited and based mainly on comparisons of mixtures with high clover contents. For example, Harkess (1963) gave mixtures containing 0, 330, 680 and 1000 g white clover per kg, and it might be inferred from his results that the digestibility benefit is likely to be slight if the clover content of the herbage is less than 300 g. $kg^{-1}$.

## Frequency of nitrogen application

In most of the experiments on the effects of nitrogen on herbage yields mentioned above the fertilizer was applied in equal split dressings at intervals over the season. Alterations in the timing of and the amounts applied at each dressing have relatively small effects on total yield per year, but may alter considerably the seasonal distribution of herbage production. Using the results from two experiments at this Institute (D. Reid, unpublished results), models of these effects have been constructed for an S23 perennial ryegrass sward cut five times annually. Fig. 7 shows the seasonal production patterns predicted from these models for two contrasting nitrogen application schedules, both supplying 335 kg nitrogen per ha annually. Schedule A is that used in most of the above-mentioned experiments when the total nitrogen rate is split into five equal dressings, one in the spring and the remainder after each cut but the last. This schedule merely accentuates the normal pattern of herbage production with a peak in June. Schedule B would be appropriate where the bulk of the herbage production is required early in the season for silage making, and thus heavier dressings are supplied for the early cuts — particularly the first — and none is applied in the late summer. This gives almost the same total yield of dry matter as schedule A, but 0.77 of it has been obtained by the 3rd week in July.

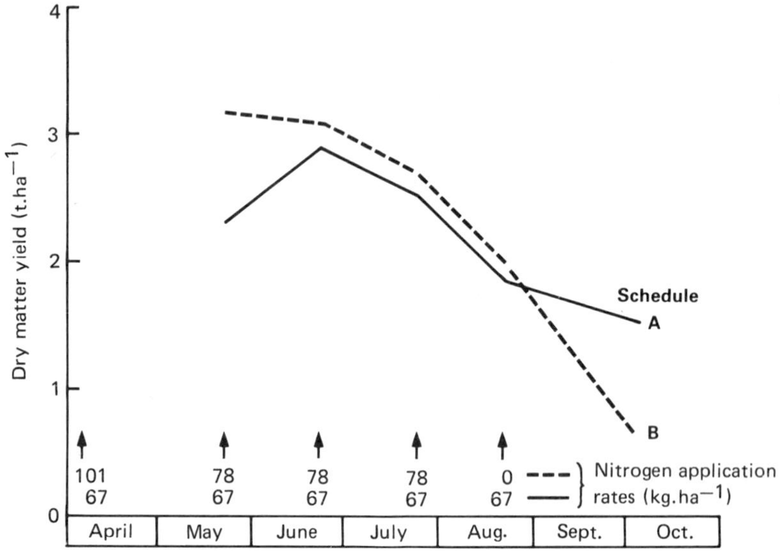


**Fig.7. Seasonal patterns of dry matter yield production from two nitrogen application schedules on S23 perennial ryegrass**

## CONCLUSIONS

The ultimate objective of studies on systems for the production of herbage for silage making should be the construction of models to predict the best system for a particular set of conditions. However, the models that may be constructed from the information available at present will, at best, describe the relationship between herbage production and only two controlling factors. An example is the model previously described (page 53), relating total production per year to nitrogen rate and frequency of cutting (Reid, 1978). However, this model gives no information on the distribution over the sequential cuts in the season. This is possible with the models recently reported by Edelsten & Corrall (1979) for the prediction of yields and digestibilities of herbage cut in different sequences of harvests, but with the nitrogen rate held constant. Until the interactions between many of the factors controlling herbage production have been characterized the construction of generalized models will be difficult if not impossible. However, the information presented in this chapter will allow the qualitative formulation of a system for the production of herbage for high-digestibility silage and the main features of this system are summarized below.

The ideal grass to include in a sward for silage making is perennial ryegrass, preferably a late variety. If nitrogen fertilizer economy is an important consideration, a large-leaved variety of white clover should be included in the sward. With a late variety of ryegrass the first cut in the season should be delayed until approximately 1 week before the date of 0.50 ear emergence. Subsequent crops should be cut at intervals of approximately 4 weeks or when the herbage is approximately 150-200 mm tall. The yield of herbage can be increased by cutting each crop as close to ground level as possible, say 25 mm, but care must be taken to avoid damaging the sward and the mower, and contaminating the cut herbage with soil. To ensure high yields of high-digestibility herbage in the early part of the season, when conditions are most suitable for silage making, the greater part of the total nitrogen dressing should be applied in the spring and after the first few cuts. The actual amounts supplied will depend on whether or not clover is included in the sward and on the nitrogen status of the soil.

## REFERENCES

ALDRICH, D.T.A. 1974 The legume — a reappraisal of its place in today's farming. *Journal of the British Grassland Society* **29** 249-251

BINNIE, R.C. & HARRINGTON, F.J. 1972 The effect of cutting height and cutting frequency on the productivity of an Italian ryegrass sward. *Journal of the British Grassland Society* **27** 177-182

BLACK, W.J.M. & ALEXANDER, J.R.B. 1967 The effects of type of cutter and height of cutting on the recovery of four grass varieties. *Journal of the British Grassland Society* **22** 260-263

CASTLE, M.E. & WATSON, J.N. 1971 Silage and milk production. A comparison between a diploid and a tetraploid ryegrass. *Journal of the British Grassland Society* **26** 265-270

CASTLE, M.E. & WATSON, J.N. 1974 Red clover silage for milk production. *Journal of the British Grassland Society* **29** 101-108

CHESTNUTT, D.M.B., MURDOCH, J.C., HARRINGTON, F.J. & BINNIE, R.C. 1977 The effect of cutting frequency and applied nitrogen on production and digestibility of perennial ryegrass. *Journal of the British Grassland Society* **32** 177-183

COLLINS, D.P. & McCARRICK, R.B. 1969 Effect of time and frequency of cutting on total and seasonal production of herbage. *Irish Journal of Agricultural Research* **8** 29-40

CORRALL, A.J., LAVENDER, R.H.& TERRY, CORA P. 1979 Grass species and varieties. Seasonal patterns of production and relationships between yield, quality and date of first harvest. *Technical Report, Grassland Research Institute* No. 26

DICKINSON, J.P. 1972 The performance and certain husbandry aspects of tetraploid broad red clovers. *Journal of the British Grassland Society* **27** 200 (Abstr.)

EDELSTEN, P.R. & CORRALL, A.J. 1979 Regression models to predict herbage production and digestibility in a non-regular sequence of cuts. *Journal of Agricultural Science* **92** 575-585

FRAME, J. 1965 The effect of winter grazing by sheep on spring and early summer pasture production. *Experimental Record, West of Scotland Agricultural College* No. 8

FRAME, J. 1975 Productivity of tetraploid red clover. *Journal of the British Grassland Society* **30** 89 (Abstr.)

GREEN, J.O., CORRALL, A.J. & TERRY, R.A. 1971 Grass species and varieties. Relationships between stage of growth, yield and forage quality. *Technical Report, Grassland Research Institute* No. 8

HARKESS, R.D. 1963 Studies in herbage digestibility. *Journal of the British Grassland Society* **18** 62-68

HUNT, I.V., FRAME, J. & HARKESS, R.D. 1975 Potential productivity of red clover varieties in S.W. Scotland. *Journal of the British Grassland Society* **30** 209-216

JONES, D.I.H. 1970 The effect of nitrogen fertilizers on the ensiling characteristics of perennial ryegrass and cocksfoot. *Journal of Agricultural Science* **75** 517-521

JONES, D.I.H., GRIFFITH, G. ap & WALTERS, R.J.K. 1961 The effect of nitrogen fertilizer on the water-soluble carbohydrate content of perennial ryegrass and cocksfoot. *Journal of the British Grassland Society* **16** 272-275

LEE, G.R., DAVIES, L.H., ARMITAGE, E.R. & HOOD, A.E.M. 1977 The effects of rates of nitrogen application on seven perennial ryegrass varieties. *Journal of the British Grassland Society* **32** 83-87

McBRATNEY, J. & LAIDLAW, A.S. 1974 The suitability of various grasses, cutting intervals and nitrogen levels for conservation systems as judged by dry matter production. *Annual Report 47, 1973-74, Agricultural Research Institute of Northern Ireland* pp 21-26

MINSON, D.J., RAYMOND, W.F. & HARRIS, C.E. 1960 Studies in the digestibility of herbage. VIII. The digestibility of S 37 cocksfoot, S 23 ryegrass and S 24 ryegrass. *Journal of the British Grassland Society* **15** 174-180

REID, D. 1959 Studies on the cutting management of grass-clover swards. I. The effect of varying the closeness of cutting on the yields from an established grass-clover sward. *Journal of Agricultural Science* **53** 299-312

REID, D. 1962 Studies on the cutting management of grass-clover swards. III. The effects of prolonged close and lax cutting on herbage yields and quality. *Journal of Agricultural Science* **59** 359-368

REID, D. 1966 Studies on the cutting management of grass-clover swards. IV. The effects of close and lax cutting on the yield of herbage from swards cut at different frequencies. *Journal of Agricultural Science* **66** 101-106

REID, D. 1970 The effects of a wide range of nitrogen application rates on the yields from a perennial ryegrass sward with and without white clover. *Journal of Agricultural Science* **74** 227-240

REID, D. 1972 The effects of the long-term application of a wide range of nitrogen rates on the yields from perennial ryegrass swards with and without white clover. *Journal of Agricultural Science* **79** 291-301

REID, D. 1977 The role of white clover in grassland production. *Report 1976, Hannah Research Institute* pp 66-69

REID, D. 1978 The effects of frequency of defoliation on the yield response of a perennial ryegrass sward to a wide range of nitrogen application rates. *Journal of Agricultural Science* **90** 447-457

REID, D. & MacLUSKY, D.S. 1960 Studies on the cutting management of grass-clover swards. Part II. The effects of close cutting with either a gang mower or a reciprocating-knife mower on the yields from an established grass-clover sward. *Journal of Agricultural Science* **54** 158-165

REID, D. & STRACHAN, N.H. 1974 The effects of a wide range of nitrogen rates on some chemical constituents of the herbage from perennial ryegrass swards with and without white clover. *Journal of Agricultural Science* **83** 393-401

THOMSON, D.J. & RAYMOND, W.F. 1970 White clover in animal production. Review of nutritional factors. In *British Grassland Society, Occasional Symposium* No. 6, pp 277-284

WAITE, R. 1958 The water-soluble carbohydrates of grasses. IV. — The effect of different levels of fertilizer treatment. *Journal of the Science of Food and Agriculture* **9** 39-43

# Chapter 4

# Silage as a foodstuff

**P.C. Thomas & D.G. Chamberlain**

*Hannah Research Institute, Ayr KA6 5HL*

The usefulness of a silage as a foodstuff depends on the weight of silage an animal will eat and on the nutritional value of the material per unit of weight. The latter is determined by the nutrient content of the silage, by the availability of the nutrients and by the efficiency with which they are used by the animal for maintenance, growth, lactation and pregnancy. For most foodstuffs, the limits of intake and nutritional value may be broadly assessed from a knowledge of the food's content of the main carbohydrate, lipid and protein components (Agricultural Research Council, 1965, 1976 & 1980; Ministry of Agriculture, Fisheries and Food *et al.*, 1975). This provides a basis for ration formulation and, to a degree, allows prediction of animal performance. With silages, however, the intake and nutritional value may vary not only with the content of the main chemical components but also with the characteristics of the silage fermentation and the silage's content of fermentation end-products. A classification of silages into five types on the basis of fermentation pattern has been devised (table 1), but it has yet to be demonstrated that the classes have sufficiently distinctive nutritional characteristics for classification to be useful in more than a qualitative sense. Most recent studies in British research institutes and universities have been, in fact, with a relatively narrow range of silages. They have generally been 'well-preserved' materials prepared from direct-cut or wilted grass using either a natural, lactic fermentation or additives containing formic acid, sulphuric acid or formaldehyde, alone or in combination. Work at the Hannah Research Institute (HRI), for example, has been predominantly with formic acid silages. This limitation on the range of materials studied is probably not serious when viewed on a national scale since a large proportion of farm silages are similar to those investigated. However, there are other types of silage, and some encountered in the field may not conform with generalizations made in this chapter.

The many experiments conducted over the years to measure food intake, growth rate and milk production in cows given silage diets, and to examine the importance of crop ensiled, harvesting method, ensilage procedure and silage additive have shown that ensilage invariably reduces the feeding value of a forage, partly through effects on food intake and partly through effects on nutrient digestion and utilization (see Wilkins, 1974 & 1978; Tayler & Wilkins, 1976; Vetter & Von Glan, 1978; Marsh, 1979; Thomas *et al.*, 1980d). Reductions in feeding value are especially marked with badly fermented 'clostridial' silages characterized by high concentrations of butyric acid and ammonia, but they also occur to a greater or lesser extent with silages judged by conventional compositional analysis to have a 'good' fermentation (see McCullough, 1978). The factors that influence the intake, digestion and utilization of silages are considered below.

**Table 1. Typical compositions of grass silages produced by five different types of fermentation (McDonald & Edwards, 1976)**

| | Silage type | | | | |
|---|---|---|---|---|---|
| | **Lactate** | **Butyrate** | **Acetate** | **Wilted** | **Chemically restricted‡** |
| pH | 3.9 | 5.2 | 4.8 | 4.2 | 5.1 |
| Dry matter (DM) (g.kg$^{-1}$) | 190 | 170 | 176 | 308 | 212 |
| Buffering capacity (mequivalents.kg$^{-1}$ DM) | 1120 | —† | 1090 | 890 | 560 |
| Protein nitrogen (g.kg$^{-1}$ total N) | 235 | 353 | 440 | 289 | 740 |
| Ammonia-N (g.kg$^{-1}$ total N) | 78 | 246 | 128 | 83 | 30 |
| Lactic acid (g.kg$^{-1}$ DM) | 102 | 1 | 34 | 59 | 26 |
| Acetic acid (g.kg$^{-1}$ DM) | 36 | 24 | 97 | 24 | 10 |
| Butyric acid (g.kg$^{-1}$ DM) | 1 | 35 | 2 | 1 | 1 |
| Water-soluble carbohydrates (g.kg$^{-1}$ DM) | 10 | 6 | 3 | 48 | 133 |
| Mannitol (g.kg$^{-1}$ DM) | 41 | — | 2 | 36 | — |
| Ethanol (g.kg$^{-1}$ DM) | 12 | — | 8 | 6 | 4 |

† Not determined
‡ Treated with a mixture of 750 g formalin: 250 g formic acid per kg at rate of 10 g. kg$^{-1}$

## SILAGE INTAKE

In all species of animals, voluntary food intake (the amount of food eaten when feeding is *ad lib.*) is regulated centrally through the activity of the hypothalamus. This small area on the lower surface of the brain acts to integrate information on the animal's requirements for nutrients, in particular for energy, with information that the animal derives about its food. The latter is obtained through the action of central nervous system receptors sited in the mouth and nose, in the digestive tract, in liver, brain and elsewhere in the body. These receptors respond to the sensory qualities of foods (taste, smell, texture etc.), to the physical effects of food ingestion on the gut (stretch, pressure etc.), to chemical stimuli arising from the end-products of digestion before and after their absorption, and to any intake-depressing compounds present in the food (see Forbes, 1980). For ruminant animals the range of foods that may be given is wide, varying from bulky low-energy forages to high-energy cereal concentrates, and the dominant food characteristic influencing intake varies with the nature of the food. Thus, for low-energy forages (e.g. hays and straws), intake is regulated predominantly by physical factors, principally the physical fill in the rumen, whilst for high energy forages and concentrates intake is regulated chemically, principally through the animal's perception of the short-chain fatty acids produced by ruminal fermentation. The relationship between food intake and the energy content of the diet is summarized schematically in fig. 1. For experimental diets consisting of cereal concentrates with various proportions of indigestible 'diluents', Baumgardt (1970) concluded that the point of change from 'physical' to 'metabolic' regulation of intake occurred at a dietary metabolizable energy (ME) concentration of approximately 9 MJ. $kg^{-1}$ dry matter (DM). However, in practice with natural diets, the amounts of food eaten, and the detail of the shape of the relationship between food intake and dietary energy concentration, will vary with the animal's energy requirements and with the particular diets under consideration. Factors such as nitrogen content, physical form, presence of intake-depressing compounds etc. are all important, as indicated in fig. 1.

Recognition that the intake of silage is lower than that of the grass from which it is made has focused attention on the products of silage fermentation to explain the reduction in intake and the differences in intake between silages. Several studies have related intake to the chemical composition of silage (see Wilkins *et al.*, 1971 & 1978) and there have been statistically significant correlations with silage pH, with the concentration of acids in the silage DM (negative correlations) and with indices of 'fermentation quality'. The latter include the proportion of ammonia-N in the total N (negative), the proportion of lactic acid in the total acids (positive), and the Flieg index (positive), which is calculated from the relative proportions of acetic, butyric and lactic acids (Zimmer, 1966). Statistical studies of this type provide useful information but suffer the shortcoming of autocorrelation between silage constituents; whilst low intake may be associated with a particular type of silage

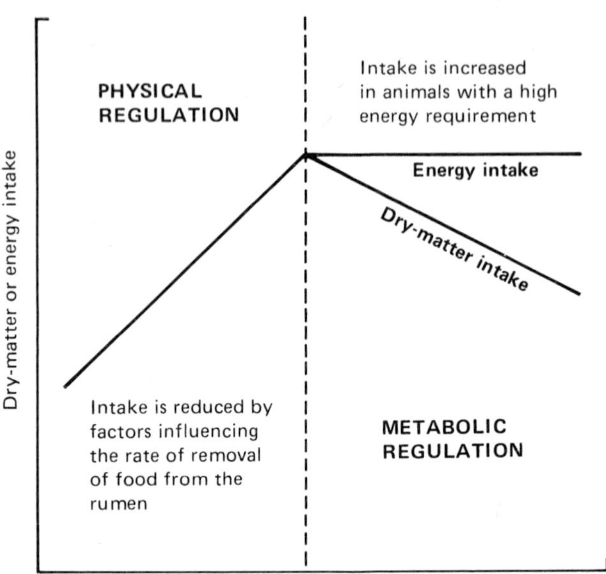


**Fig.1. A schematic representation of the relationship between voluntary food intake in the cow and the concentration of energy in the diet.** The diagram shows the change that occurs as the dominant mechanism for intake regulation changes from a physical to a metabolic type. Where intake regulation is physical, energy intake and dry-matter intake both increase with the energy content of the diet. Under these conditions also, intake is reduced by physical characteristics of the diet (e.g. increase in chop-length), and by nutritional and physiological factors (e.g. low dietary protein contents, high dietary starch contents and low efficiency of rumination) that reduce the rate of removal of plant fibre constituents from the rumen. Where intake regulation is metabolic, energy intake remains constant as dietary energy concentration is increased, and dry-matter intake is reduced. At all stages of the relationship intake tends to be increased in animals with a high energy requirement

fermentation it is not possible in this type of study to ascribe a causative intake-depressing action to specific silage components. Direct investigations of the effects of individual fermentation products have been undertaken, but they have not allowed clear-cut conclusions about the agents influencing silage intake or about the regulatory mechanisms involved. Partial neutralization of silage with sodium bicarbonate has increased intake in some experiments but not in others (table 2).

**Table 2. The effect of partial neutralization of silages with sodium bicarbonate on the voluntary intake of calves, sheep and cows**

| Type of silage | Test animals | Addition to silage | pH of the silage at feeding | Organic matter intake (g. (kg $W^{0.75})^{-1}.d^{-1}$) | Response to $NaHCO_3$ (proportion of control intake) | Source of data‡ |
|---|---|---|---|---|---|---|
| Unwilted grass silage made without additive | Calves | None | 4.0 | 66.2 | | |
| | | $NaHCO_3$ | 5.5 | 78.0 | +0.151 | 1 |
| Unwilted grass silage made without additive | Calves | None | 4.1 | 80.0 | | |
| | | $NaHCO_3$ | 5.4 | 84.4 | +0.055 | |
| Maize silage | Calves | None | 3.9 | 68.9 | | |
| | | $NaHCO_3$ | 5.4 | 70.5 | +0.022 | 2§ |
| Maize silage made with added urea (20 g.kg$^{-1}$) | Calves | None | 4.0 | 70.5 | | |
| | | $NaHCO_3$ | 5.5 | 79.7 | +0.130 | |
| Unwilted grass silage made without additive | Sheep | None | 4.1 | 51.3 | | |
| | | $NaHCO_3$ | 5.3 | 54.6 | +0.064 | 1 |
| Unwilted grass silage made without additive | Sheep | None | 4.1 | 47.5 | | |
| | | $NaHCO_3$ | 5.4 | 52.4 | +0.103 | 1 |
| Unwilted grass silage made without additive | Sheep | None | 3.9 | 47.6† | | |
| | | $NaHCO_3$ | 6.1 | 46.3 | −0.027† | 3 |
| Unwilted grass silage made without additive | Sheep | None | 4.1 | 43.3† | | |
| | | $NaHCO_3$ | 7.7 | 47.6 | +0.099† | 3 |
| Wilted grass silage made with formic acid as additive | Sheep | None | 3.8 | 44.3 | | |
| | | $NaHCO_3$ | 4.2 | 46.4 | +0.047 | 4 |
| | | $NaHCO_3$ | 5.0 | 46.6 | +0.052 | |
| Unwilted grass silage made without additive | Cows | None | 3.9 | 70.3† | | |
| | | $NaHCO_3$ | 7.0 | 63.5 | −0.097† | 3‖ |
| Wilted grass silage made with formic acid as additive | Cows | None | 3.8 | 81.5 | | |
| | | $NaHCO_3$ | 4.3 | 75.7 | −0.075 | 4 |
| | | $NaHCO_3$ | 4.9 | 69.3 | −0.149 | |

† Values are DM intake
‡ 1, McLeod *et al.* (1970); 2, Thomas & Wilkinson (1975); 3, Lancaster & Wilson (1975); 4, Farhan & Thomas (1978)
§ For presentation of results data have been normalized to a live weight of 100 kg
‖ For presentation of results the Jersey cows used have been assumed to have an average live weight of 350 kg

Responses to bicarbonate may be affected by a variety of factors, including the species and age of the animals under consideration (calves seem to be particularly responsive), the type of silage and the level of bicarbonate supplementation. Responses are most likely to occur in animals suffering acid stress but, contrary to what is often believed, this condition is not an inevitable consequence of the consumption of low-pH silage. Dietary additions of lactic acid have reduced silage intake in several experiments, but effects are not invariably obtained, and appear to be offset if the diet is supplemented with a protein food (McLeod *et al.*, 1970; Thomas *et al.*, 1980a). Dietary supplements of acetic acid (Hutchinson & Wilkins, 1971) or of nitrogenous fermentation products, such as ammonia, histamine, tryptamine and tyramine, have generally not reduced silage intake significantly (see Vetter & Von Glan, 1978).

A number of recent studies with highly digestible, well preserved silages have highlighted the importance of physical, 'rumen-fill' factors in the regulation of silage intake. In cows given ryegrass silages, reduction of rumen volume through the insertion of water-filled bags into the rumen has reduced intake by 16.5% (Farhan & Thomas, 1978), whilst in sheep intake has been increased by up to 19% by fine mincing of silage before feeding (Dulphy & Demarquilly, 1973; Thomas *et al.*, 1976b). From early studies, Campling (1966) concluded that particles of silage were retained in the rumen substantially longer than those of hay, suggesting that the ruminal breakdown of silage was slow. Incubation of forages in small nylon bags in the rumen has shown that the activity of the rumen cellulolytic bacteria is not reduced by silage feeding, and there is little difference in the rate of chemical breakdown in the rumen between fresh grass and the same material conserved as silage or hay (Deswysen *et al.*, 1978; Smith *et al.*, 1980). In contrast, there is now considerable evidence to indicate that the eating and ruminating behaviour of animals given silages is different from that of animals given fresh or dried forages. When given silages, animals spend less time per day eating and eat a greater number of small meals. The time spent eating and ruminating per kg DM intake is increased, and there is a longer 'inactivity' period between the completion of a meal and the start of rumination; the incidence of pseudo-rumination, i.e. rumination reflex not accompanied by the regurgitation of a digesta bolus, is also increased (Dulphy & Demarquilly, 1973; Deswysen *et al.*, 1978; Castle *et al.*, 1979). The cause(s) of the behavioural pattern observed with silage diets has not yet been fully elucidated but Clancy *et al.* (1977) made an important observation when they showed that intraruminal infusion of lucerne silage juice reduced rumen motility and the rate of eating in sheep. This finding has now been confirmed and extended in studies with grass silages prepared with formic acid as an additive (Smith & Clapperton, 1981). The implication is that the feeding behaviour of animals given silage diets is influenced by the presence of some as yet unidentified compound(s), possibly amine(s), present in silage juice (Clancy *et al.*, 1977; Thomas *et al.*, 1980d). Thus it may be that the chemical end-products of silage fermentation affect silage intake through a 'physical' mechanism of intake regulation.

Irrespective of whether this is the case or not, however, silage intake will be influenced by dietary factors that affect rumen-fill, including the rate of digestion of silage in the rumen and the rate of passage of silage particles to the lower gut, and the differing effects of supplementary foods on silage intake described in Chapter 6 may be explained largely in these terms. Coarse bulky foods, such as chopped hay, reduce the available rumen volume and act as a substitute rather than as a true supplement for silage. Concentrated foods, on the other hand, have only a small effect on rumen volume, but foods rich in starch or sugars reduce the activity of cellulose- and hemicellulose-digesting bacteria in the rumen, and slow the breakdown of silage fibre, whilst those rich in protein have the opposite effect (Morgan *et al.*, 1980; Thomas *et al.*, 1980f).

## SILAGE ENERGY AND PROTEIN

### General comments

Following its proposal in 1965, the Metabolizable Energy System has become the standard for the energy rationing of ruminant livestock in the United Kingdom. The system has been discussed fully elsewhere (ARC, 1965 & 1980; MAFF *et al.*, 1975) but, briefly, it depends on the estimation or computation of dietary ME intake, where

ME intake = gross energy (GE) intake − faecal energy loss − urine energy loss − methane energy loss,

together with an estimate of the efficiency ($k$) with which ME is used to meet the animal's requirements for maintenance ($k_m$), fattening ($k_f$) and lactation ($k_{lo}$). The $k$ factors in the system are based on the results of calorimetric experiments that have been used to establish empirical relationships between the efficiences of utilization of ME and the dietary ME concentration (metabolizability ($Q_m$)). The efficiency factors vary with the type of diet and with the physiological processes (maintenance, fattening etc.) under consideration. The causes of these variations have not been fully resolved, but they are related to the energetics of the digestion processes and to the composition of the mixture of end-products of digestion that are absorbed from the gut and metabolized in the tissues (see Thomas & Rook, 1977; Blaxter, 1980; Webster, 1980). The main end-products are: acetic, propionic and butyric acids formed by bacterial and protozoal fermentation of food constituents in the rumen, and to a lesser extent in the caecum and colon; and glucose, amino acids and long-chain fatty acids absorbed in the small intestine, and derived partly from dietary and partly from microbial sources (see fig. 2).

Dietary protein requirements and allowances have traditionally been expressed in terms of digestible crude protein (DCP) (digestible nitrogen × 6.25) and this is still a common approach. Recently, however, a new system for protein rationing has been devised that takes account of the fact that the supply of amino acids to the animal's tissues is a function of the amount and composition of the protein digested in the small intestine, and that this in turn depends on the amount of dietary protein surviving rumen fermentation undegraded, and on the protein synthesized by the rumen microorganisms (fig. 2). The system has been put forward by the ARC (1980) 'as a framework for future research efforts and as a means of focussing attention on factors for which additional data are required' but, despite its acknowledged limitations, the system has been adopted widely in commerce and is in practical day to day use for diet formulation.

## Silage energy

### *Gross energy*

The laboratory analysis of silage presents special problems because of the volatile compounds present. Thus silage DM content is underestimated by oven-drying procedures, and satisfactory determination requires a toluene distillation technique in which both moisture and volatiles are taken into account (Dewar & McDonald, 1961). Similar problems arise in the determination of GE content and, for accurate values, bomb calorimetry measurements must be made on fresh silage samples (Nijkamp, 1965; McDonald *et al.*, 1973). A number of authors have drawn attention to the fact that silages frequently have GE contents higher than those observed with other forages, although some of the values in the literature are misleading since they reflect determinations of GE on fresh silage but expression of the results per kg of oven-dry matter. McDonald *et al.* (1973) have shown that GE content will be elevated by heterolactic silage fermentation yielding end-products of

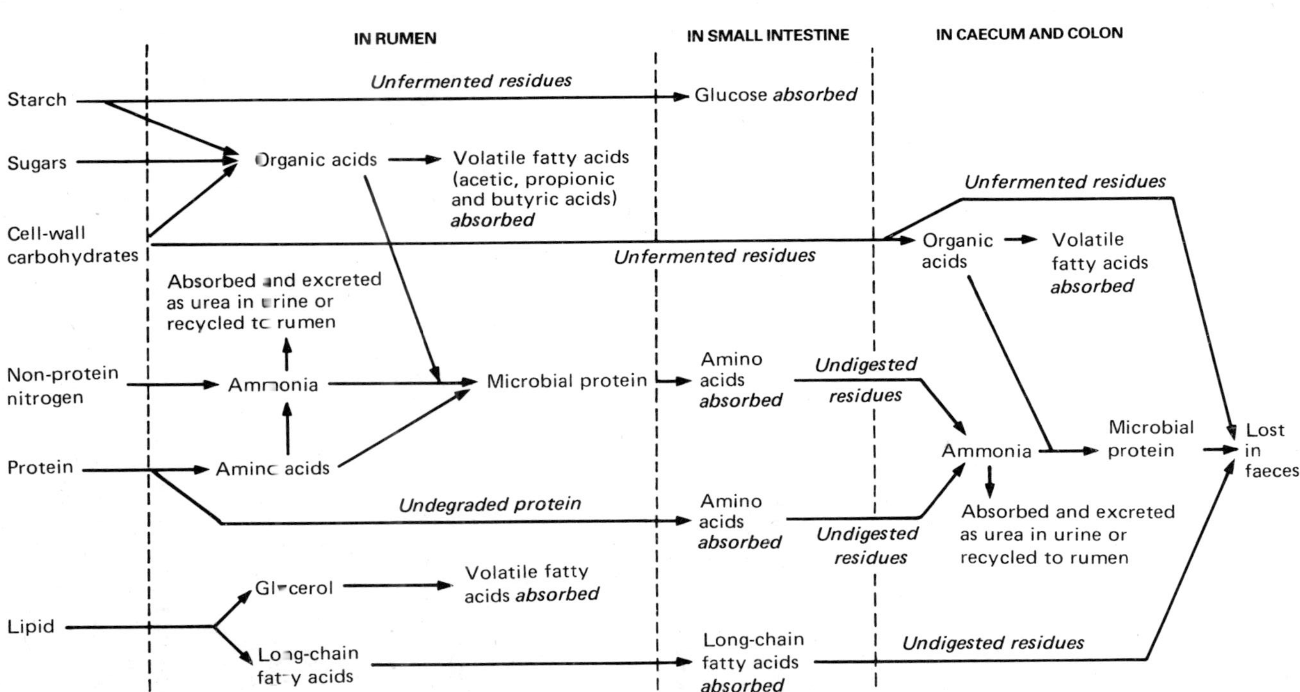


**Fig.2.** **A schematic representation of digestion of dietary constituents in the ruminant, showing the main products of digestion in the rumen, small intestine, caecum and colon, and their fates**

high calorific value. With a group of 36 medium- and low-digestibility grass and legume silages varying widely in method of preparation and in composition, Terry & Osbourn (1980) found that the GE content of the organic matter was, on average, 7.5% higher than that observed with fresh forages. The mean GE value obtained for the silages was 18.88 (range 17.39-19.89) MJ. $kg^{-1}$ DM, and similar values have been obtained for silages of high digestibility. For nine ryegrass silages made with formic acid at the HRI mean energy content was 18.59 (17.47-19.98) MJ. $kg^{-1}$ DM, and for 16 grass and grass-clover silages of various types obtained from commercial farms for experiments at the Rowett Research Institute (RRI) values were 19.03 (18.46-19.50) MJ. $kg^{-1}$ DM (Wainman *et al.*, 1979).

### *Digestible organic matter and digestible energy*

Substantial reductions in the digestibility of the organic matter in grass may occur during ensilage if there is extensive oxidative fermentation or high effluent loss but, if ensilage conditions are good, changes in the digestibility are small, except when formalin is used as an additive, when there may be a reduction of 0.02-0.04 in digestibility coefficient (see Wilkins, 1974). The digestible organic matter content of silage dry matter (DOMD) therefore reflects that of the original grass and is influenced by the grass composition, and particularly by its stage of maturity at harvest, as discussed in earlier chapters.

In a number of experiments, data on the digestibility of organic matter and of GE have been obtained simultaneously, and the relationship between DOMD and digestible energy (DE) may be examined. In the work of Terry & Osbourn (1980), referred to above, the silages ranged in DOMD value from 498 to 632 g. $kg^{-1}$ and DE (MJ. $kg^{-1}$ DM) varied with DOMD and crude protein (CP) concentration (g. $kg^{-1}$ DM) according to the equation:

$$DE = 0.02136\,DOMD + 0.041\,CP - 0.000103\,(CP)^2 - 4.22;$$

a corresponding relationship,

$$DE = \frac{DOMD}{1000}(12.99 + 0.0432\,CP),$$

was also calculated in earlier work by Kelly & Thomas (1978a) for a small group of ryegrass silages of 650-700 g. $kg^{-1}$ DOMD. Equations including crude protein as a variable may not always be convenient, however, and other relationships between DE and DOMD may be proposed. These are most accurate if they take account of the GE value (MJ. $kg^{-1}$) of the silage organic matter and an equation

$$DE = 0.01978\ DOMD + 0.574\ GE - 11.6$$

has been suggested by Terry & Osbourn (1980). Except for silages with a substantial content of high-energy fermentation products, however, DE may be calculated with reasonable precision without correcting for GE content. For the silages they examined, Terry & Osbourn (1980) found

$$DE = 0.02366\ DOMD - 2.133\ (no. = 36;\ r^2 = 90.6),$$

whilst analysis of data for a narrower range of ryegrass silages of 650-700 g. kg$^{-1}$ DOMD used at the HRI gave

$$DE = 0.03824\ DOMD - 12.693\ (no. = 8;\ r^2 = 81.9),$$

and recalculation of results for commercial silages of 640-780 g. kg$^{-1}$ DOMD used in experiments at the RRI (Wainman *et al.*, 1979) gave

$$DE = 0.01902\ DOMD + 0.542\ (no. = 16;\ r^2 = 79.4).$$

For silages of high digestibility (*ca.* 700 g. kg$^{-1}$) these three equations are in reasonable agreement but, because of the limits of the range of data on which the equations are based, they become increasingly divergent at low DOMD. An alternative approach is to calculate the mean energy value per kg of digestible organic matter (DOM) (see Swift, 1957). For the silages examined by Terry & Osbourn (1980) the GE of the DOM was related to the crude protein (CP) content by the equation

$$GE = 13.98 + 0.072\ CP - 0.000184\ (CP)^2,$$

which, for typical grass silages containing 140-180 g crude protein per kg DM, indicates energy values of 20.3-20.9 MJ. kg$^{-1}$ DOM. Corresponding, but directly calculated, average values for the higher-digestibility silages used at the HRI and RRI are 19.71 (s.e. 0.47) and 19.07 (s.e. 0.10) respectively, and the mean for this range of silages is 19.35 (s.e. 0.18) MJ. kg$^{-1}$ DOM.

## *Metabolizable energy*

For rationing purposes the ME of forages is commonly calculated from the estimated DE content using the factor 0.81, to allow for the DE loss in methane and urine (see Morgan & Barber, 1980). This factor is based on calorimetric experiments with dried foods and on data obtained in initial studies at the HRI in the early 1960s, with silages of low or moderate digestibility (see ARC, 1965). In the past few years more data on high-digestibility silages, especially those made with formic acid additives, have been obtained and the evidence now suggests that ME/DE for this type of silage is higher than allowed for by the conversion factor currently in use. For six silages given without supplements in experiments at the HRI, ME/DE was 0.844 (s.e. 0.004), whilst calculations based on data for 16 silages studied at the RRI (where the experimental approach was to calculate the values for ME and DE from experiments with diets containing various proportions of silage with wheat or wheat offals) gave a value of 0.838 (s.e. 0.002). On this basis an ME/DE of 0.84 would seem appropriate for use in calculations of ME.

Several methods have been published for the estimation of DOMD in forages using *in vitro* digestibility techniques, or correlations of digestibility with chemically determined components such as fibre or lignin (Tilley & Terry, 1963; Alexander, 1969; Morrison, 1973; Jones & Hayward, 1975; Morgan & Barber, 1980), and these methods may provide valid estimates of the DOMD of silages provided that they are arranged to take account of the volatile components in the calculation of both the DOM and DM terms. Determination of *in vitro* digestibility on dried material and expression of the results on an oven-dry matter basis will underestimate the true DOMD by an amount increasing with the proportion of volatile compounds in the DM; typically, for well preserved silages, the error is 20-40 g. $kg^{-1}$. For most dried foods ME (MJ. $kg^{-1}$ DM) may be computed from 0.015 $\times$ DOMD, an expression which assumes that the energy content of DOM is 19.0 MJ. $kg^{-1}$ and that ME/DE is 0.81. For silages, higher coefficients of 0.017-0.018 have been suggested to allow for the high calorific value of the DOM (see Morgan & Barber, 1980). On the basis of the present evidence, however, for well preserved high-digestibility silages an appropriate coefficient is 0.016 (19.35 MJ. $kg^{-1}$ $\times$ 0.84).

### *Efficiency of utilization of metabolizable energy*

There have been relatively few studies of the efficiency of utilization of silage ME and the work undertaken to date has been with a narrow range of silages. Values for $k_m$ and $k_f$ determined in sheep or predicted from the equation given by the ARC (1965 & 1980) are given in table 3. There is quite good accordance between the determined and calculated values of $k_m$ (fig. 3), despite the fact that the prediction equations are derived from experiments with non-silage diets. The position for $k_f$ is less clear-cut, however, for, whilst for most of the diets examined divergence between observed and calculated values is not unduly large (fig. 3), for a small number of silages low $k_f$ values have been obtained. In some experiments the unusually low values for $k_f$ may be explained partly by technical problems in making accurate determinations where energy retention is limited by the amount of silage that the animal will

**Table 3.** **The partial efficiencies of utilization of the metabolizable energy (ME) of silage diets for maintenance ($k_m$) and fattening ($k_f$) determined experimentally and calculated by the equations of the ARC (1965 & 1980)**

| Diet and details of silage | Determined | | Calculated: ARC, 1965† | | Calculated: ARC, 1980‡ | | Source of data |
|---|---|---|---|---|---|---|---|
| | $k_m$ | $k_f$ | $k_m$ | $k_f$ | $k_m$ | $k_f$ | |
| *Silage alone* | | | | | | | |
| First cut perennial ryegrass, wilted and ensiled with formic acid | 0.69 | 0.21 | 0.72 | 0.50 | 0.75 | 0.44 | Kelly & Thomas (1978a) |
| Regrowth perennial ryegrass, unwilted and ensiled with formic acid | 0.71 | 0.57 | 0.75 | 0.57 | 0.74 | 0.46(0.56)‡ | |
| Regrowth perennial ryegrass, unwilted and ensiled with formic acid | 0.68 | 0.59 | 0.71 | 0.48 | 0.70 | 0.34(0.42)‡ | |
| First cut perennial ryegrass, wilted and ensiled with formic acid | — | 0.51 | — | 0.57 | — | 0.56 | N.C. Kelly & P.C. Thomas (unpublished results) |
| First cut grass, wilted | 0.66 | 0.42 | 0.71 | 0.48 | 0.70 | 0.42 | Ekern & Sundstøl (1974) |
| First cut grass, wilted and ensiled with formic acid | 0.71 | 0.41 | 0.72 | 0.50 | 0.71 | 0.45 | |
| First cut grass | – | 0.49 | – | 0.49 | – | 0.43 | Smith *et al.* (1977) |
| First cut grass | – | 0.54 | – | 0.54 | – | 0.51 | |
| Third cut grass | – | 0.51 | – | 0.48 | – | 0.34(0.43)‡ | |
| First cut grass wilted and ensiled with formic acid | 0.68 | 0.27 | 0.72 | 0.49 | 0.70 | 0.43 | Sundstøl *et al.* (1980) |
| *Mixed silage and concentrates* | | | | | | | |
| First cut perennial ryegrass, wilted and ensiled with formic acid plus rolled barley (820:180 g. kg$^{-1}$)§ | – | 0.43 | – | 0.53 | – | 0.51 | Kelly & Thomas (1978a) |
| First cut perennial ryegrass, wilted and ensiled with formic acid plus rolled barley (700:300 g. kg$^{-1}$)§ | – | 0.52 | – | 0.55 | – | 0.52 | |
| Regrowth perennial ryegrass, unwilted and ensiled with formic acid plus rolled barley (630:370 g.kg$^{-1}$)§ | – | 0.54 | – | 0.54 | – | 0.52 | |
| First cut grass wilted and ensiled with formic acid plus cereal-based concentrate | 0.75 | 0.50 | 0.73 | 0.53 | 0.72 | 0.52 | Sundstøl *et al.* (1980) |

† Calculated from equations of ARC (1965): $k_m = 0.3$ ME/GE + 0.546; $k_f = 0.81$ ME/GE + 0.03, where GE is gross energy

‡ Calculated from equations of ARC (1980): $k_m = 0.35$ ME/GE + 0.503; $k_f = 1.32$ ME/GE − 0.318 (for first growth grass); $k_f = 1.16$ ME/GE − 0.308 (for regrowth grass); $k_f = 0.38$ ME/GE + 0.282 (for mixed diets). Values in parentheses are those calculated using equations for first growth grass

§ Proportions of silage:barley on a DM basis; $k_m$ value used to calculate $k_f$ has been assumed to be that given by the ARC (1965) equation

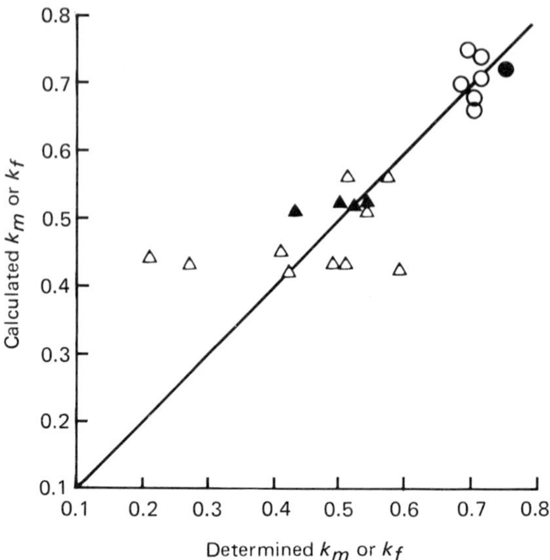


Fig.3. **The relationship between calculated and determined values for $k_m$ and $k_f$.** Values for $k_m$ have been calculated for silage (○), and silage and concentrate (●), diets using the equations of the ARC (1980). Values for $k_f$ have been calculated for silage (△), and silage and concentrate (▲), diets using the equations of the ARC (1980) with the exception that the equation for first growth grass has been used for all cuts (see table 3). The straight line represents the line of exact agreement between predicted and observed values

eat, but in other experiments this is not so and the low $k_f$ appears to result from a true inefficiency in energy utilization. It was initially suggested that the inefficiency might be associated with silages of high D (−)-lactic acid content (Kelly & Thomas, 1978a) but subsequently this has been shown to be incorrect (N.C. Kelly and P.C. Thomas, unpublished results). At present the cause of the inefficiency is unknown and the occurrence of low $k_f$ values is unpredictable. Furthermore, Kelly & Thomas (1978a) found that with a 'low $k_f$' silage the inefficiency in energy utilization was apparent only when the silage was given unsupplemented and at a high level of feeding, and thus there may be little problem under farm conditions where silages are most commonly given as part of a mixed diet.

The calorimetric efficiency of utilization of ME for lactation ($k_{lo}$) varies with diet over a relatively narrow range from 0.56-0.66 (ARC, 1980) and studies conducted with cows given mixed forage and concentrate diets have indicated that inclusion of silage as a forage has no adverse effect on the efficiency observed (Van Es & Nijkamp, 1969). The efficiency of utilization of ME in cows given diets containing a high proportion of highly-digestible silage has yet to be investigated.

## *Digestion of silage energy constituents*

In recent studies animals fitted with cannulae in the proximal duodenum and terminal ileum have been used to examine the quantitative importance of the rumen, small intestine, and caecum and colon as sites for the digestion of silage constituents. The results available show that, on average, approximately 0.63 of the digestible energy 'disappears' in the rumen and that corresponding values for the small intestine, and caecum and colon, are 0.27 and 0.10 respectively (see table 4). The figures for the digestion of organic matter follow a broadly similar pattern (table 5), although the proportion of digested organic matter disappearing in the rumen tends to be higher than the corresponding figure for energy because of the relatively high calorific value of the digestion end-products absorbed in the small intestine. For the silage carbohydrate constituents also, the rumen is the major site of digestion and, typically, 0.90 or more of the digestible cellulose and hemicellulose in the diet is digested before the small intestine. This pattern is changed substantially, however, if the diet is supplemented with cereals. Thomas *et al.* (1980f) found that, when sheep receiving 644 g silage DM per d were given a supplement of 151 g barley DM per d, the digestibility of dietary cellulose was reduced from 0.715 to 0.682 and the proportion of the digestion occurring in the rumen was reduced from 0.90 to 0.77.

Of the digestion end-products absorbed in the small intestine, amino acids account on a caloric basis for more than 0.70 and the remainder must be composed mainly of carbohydrate and lipid. There is, however, little passage of sugars or starch to the duodenum in animals given diets of silage alone; and even where the diet is supplemented with barley, less than 0.10 of the starch in the diet survives rumen fermentation (Thomas *et al.*, 1980f), so that the amount of carbohydrate absorbed from the small intestine is relatively low.

Of the large proportion of silage DE digested in the rumen most is absorbed as volatile fatty acids (VFA). Estimates based on the difference between the amounts of GE consumed and passing to the duodenum, and on methane production measured by respiration calorimetry, indi-

**Table 4. The sites of digestion of dietary energy in sheep given various diets of silage and concentrates. (Silages were from primary or regrowth cuts of perennial ryegrass or predominantly perennial ryegrass swards)**

| Details of silage | | Diet | Gross energy intake (MJ. d$^{-1}$) | Digestibility of gross energy | Proportion of digested energy disappearing in the | | | Source of data |
|---|---|---|---|---|---|---|---|---|
| Post-cutting treatment | Additive | | | | Rumen | Small intestine | Caecum and colon | |
| Direct cut | None | Silage | 18.87 | 0.720 | 0.568 | 0.339 | 0.093 | Beever *et al.* (1971) |
| Wilted | None | Silage | 16.65 | 0.675 | 0.614 | 0.274 | 0.274 | |
| Direct cut | None | Silage | 20.1 | 0.795 | 0.701 | 0.262 | 0.037 | Beever *et al.* (1977) |
| Direct cut | Formaldehyde† | Silage | 19.8 | 0.746 | 0.609 | 0.290 | 0.101 | |
| Wilted | Formic acid‡ | Silage | 11.63 | 0.655 | 0.625 | 0.241 | 0.133 | Thomas *et al.* (1980f) |
| | | Silage + barley (800:200 g. kg$^{-1}$) | 14.41 | 0.694 | 0.594 | 0.286 | 0.121 | |
| Direct cut | Formic acid‡ | Silage | 13.65 | 0.793 | 0.683 | 0.221 | 0.096 | Thomas *et al.* (1980f) |
| Direct cut | Formic acid‡ | Silage | 13.87 | 0.729 | 0.668 | 0.211 | 0.121 | |
| Direct cut | Formic acid§ | Silage | 9.09 | 0.659 | 0.621 | – | – | Unsworth & Stevenson (1978) |
| Wilted | Formic acid§ | Silage | 7.18 | 0.594 | 0.594 | – | – | |
| Direct cut | Formic acid§‖ | Silage | 10.19 | 0.734 | 0.569 | – | – | |
| Wilted | Formic acid§ | Silage | 9.82 | 0.727 | 0.583 | – | – | |
| Not stated | None | Silage‡‡ | 9.69 | 0.643 | 0.709 | – | – | |
| Direct cut | Formic acid¶ | Silage | 47.2‖‖ | 0.732 | 0.659 | – | – | Thomson *et al.* (1981) |
| | Formic acid -formaldehyde†† | Silage | 50.7 | 0.696 | 0.602 | – | – | |
| | Formic acid -formaldehyde†† | Silage + urea§§ | 43.7 | 0.703 | 0.653 | – | – | |
| | | **Mean for all diets** | | 0.706 | 0.628 | 0.266 | 0.101 | |
| | | **s.e.** | | 0.013 | 0.011 | 0.014 | 0.010 | |

† Formalin (380 g formaldehyde per l) at 6.4 l. t$^{-1}$ fresh grass
‡ Formic acid (850 g. l$^{-1}$) at 2.3 l. t$^{-1}$ fresh grass
§ Rate of application not given
‖ Urea added at ensilage at rate of 10 kg. t$^{-1}$ grass DM
¶ Formic acid (210 g. l$^{-1}$) at 7.1 l. t$^{-1}$ fresh grass
†† Formic acid (850 g. l$^{-1}$) and formalin (350 g formaldehyde per l) in equal proportions at 8.5 l. t$^{-1}$ fresh grass
‡‡ Silage made from permanent pasture
§§ Urea (20 g. kg$^{-1}$ DM) and $Na_2SO_4$ (0.21 kg. kg$^{-1}$ urea) added at feeding
‖‖ Experiments with calves

cate that rumen VFA production with silage diets accounts for 0.53 of the DE (Thomas *et al.*, 1980e). Similar or slightly lower values have been estimated directly by isotope dilution methods using $^{14}C$-labelled acids (Beever *et al.*,1977; N.C. Kelly, P.C. Thomas and N.H. Strachan, unpublished results). Fermentation in the rumen in animals given silage diets has several features of note. In animals fed twice daily the composition of the mixture of VFA present shows a pronounced change after feeding and a marked peak in the proportion of propionic acid. As a consequence, samples of rumen fluid for analysis must be taken frequently throughout the feeding cycle or at times selected carefully after feeding to provide values representative of the average composition of the VFA mixture. The cause of the peak in propionic acid production appears to be the fermentation of the lactic acid in the silage; in animals given silage, both L(+)- and D(−)-lactic acids are fermented in the rumen with a half-life of approximately 25 min (Kelly & Thomas, 1978b) and a major product of their fermentation is propionic acid. Somewhat surprisingly, the fermentation activity is located principally in the protozoal rather than bacterial fraction of rumen digesta (D.G. Chamberlain, P.C. Thomas and Fiona J. Anderson, unpublished results), and when animals are defaunated the post-feeding peak in propionate is reduced and there is an increase in the proportion of butyrate in the fermentation products (Chamberlain *et al.*, 1980).

Despite the post-feeding peak in propionic acid, however, the average composition of the mixture of VFA in the rumen does not have a high proportion of propionic acid and, in animals given diets solely of highly digestible silage, the proportions of acetic, propionic and butyric acids are typically 600-700, 160-260 and 70-110 mmol. $mol^{-1}$ of total VFA, respectively. Moreover, it has been the experience in our experiments that, even where the proportion of silage in the diet is low (as little as 200-250 g. $kg^{-1}$ of the DM) and where the remainder of the diet is cereal concentrate, or concentrate and ground cubed forage lacking in fibrousness, 'high propionate' fermentation patterns typical of those observed with dried forage and concentrate diets (see Thomas & Rook, 1977) rarely occur (see fig. 4).

## Silage protein

### *Assessment of protein value*

Equations for the calculation of DCP (g. $kg^{-1}$ DM) from dietary crude protein (CP) concentration, such as

DCP = 0.9115 CP − 36.7 (Watson & Nash, 1960),

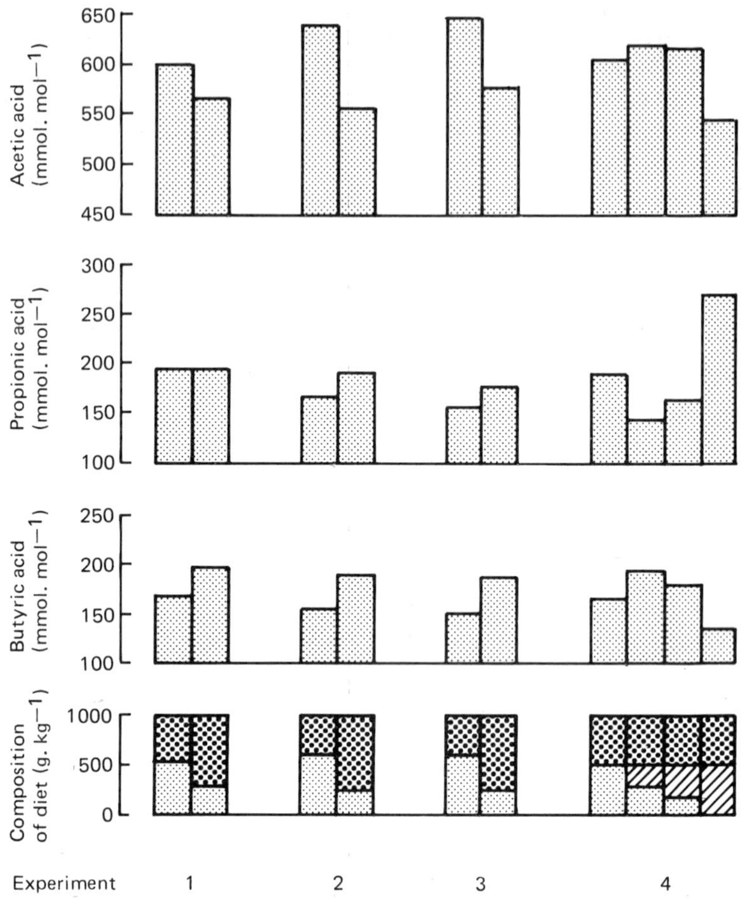


**Fig.4. The molar proportions of short-chain fatty acids in the rumen of cows given diets consisting of silage and dairy concentrates or silage, ground and pelleted dried grass, and dairy concentrates in various proportions.** Diet composition shown by proportion of silage in diets ( ▩ ), proportion of various types of dairy concentrates in diets ( ▢ ) and proportion of ground and pelleted dried grass in diets ( ▨ ). (Each histogram column represents a mean value for 4-7 cows. Data are from experiments conducted at the HRI)

were developed from the early digestibility studies with silages and have formed the basis for the assessment of silage protein for rationing purposes. However, because of uncertainty about the nutritive value of the silage non-protein N fraction, nutritionists have tended to accept the calculated silage DCP figures with caution, reducing the values either through the calculation of protein equivalents (Woodman, 1948) or through the use of arbitrary adjustments based on the maturity and quality of preservation of the silage, as judged from the silage's content of fermentation end-products. N balance studies were the first to show that the low efficiency of N utilization in animals given silage diets might be improved by supplementation of silages with cereals (see Wilkins, 1974) and, although uncertainties about the value of silage crude protein remained, the concept of 'balanced' silage-barley diets for beef and dairy cattle was widely adopted in the UK during the 1960s and early 1970s. More recently, however, the value of such diets has been questioned following demonstrations of responses in growth rate or milk production to the supplementation of silage diets with protein rich foods; indeed, good results have been obtained using oilseed meals as sole supplements (see Chapter 6).

The questions surrounding the choice of appropriate supplements for animals given silage are, of course, part of a more general problem in the protein rationing of ruminant livestock and recognition of this problem has led to the development of the new rationing system proposed by the ARC (1980). In the new system the total amino acid supply that the animal derives from its diet is calculated as the proportion of the duodenal flow of protein (amino acid-N × 6.25) that is digested and absorbed in the small intestine, and the duodenal flow of protein is given by the sum of the dietary protein that passes through the rumen undegraded and the rumen-synthesized microbial protein. In the absence of a satisfactory laboratory method for routine determination of the degradabilities of individual samples of foods, a classification of foods into four groups, varying in the degradability coefficient of protein, has been suggested. The groups are: A (0.71-0.90), B (0.51-0.70), C (0.31-0.50) and D (<0.31). For microbial protein synthesis, factors have been proposed to allow calculation from ME intake. The principle employed is to convert ME to DOM by dividing by a factor of 15.58 (which assumes ME/DE = 0.82 and DOM = 19.0 MJ DE per kg) and to multiply this value by 0.65, which is the proportion of DOM suggested to be digested in the rumen, and by 30, which is the adopted rate of rumen microbial cell synthesis (g N per kg OM apparently digested in the rumen). Of the microbial N passing to the duodenum, 0.80 is assumed to be amino acid-N, and the proportion of microbial and undigested dietary N absorbed in the small intestine and deposited in the tissues is calculated using 0.70 as the digestibility coefficient and 0.75 as the efficiency of utilization of absorbed amino acids. Thus, provided that there is no dietary deficiency restricting ruminal protein synthesis, total amino acid-N absorbed in the small intestine and deposited in the tissues is:

$$0.75 \times 0.7 \text{ (undegraded dietary N} + \text{ME } (1/15.58 \times 0.65 \times 30 \times 0.80)).$$

The present use of constant calculation factors limits the accuracy of the system since the proportion of DOM digested in the rumen, the rate of microbial synthesis etc. are known to vary with diet (see Thomas, 1973 & 1977). This may be particularly important with silage diets since the data-base from which the factors were calculated contained few results for silages.

### *Studies on the digestion of silage nitrogen*

Information on quantitative aspects of the digestion of silage N is, in fact, now accumulating and many of the data known to be available from studies in the sheep are summarized in table 6. The range of diets that has been examined is far from complete but includes a variety of types of silage that have been given either alone or with concentrate foods. As has been shown for dried forage diets, the combined effects of dietary protein degradation and microbial protein synthesis in the rumen may result in the amount of non-ammonia N (NAN) passing to the duodenum being greater or less than that ingested in the food. The tendency is for net losses of N across the rumen to occur when the ratio of N to DOM in the diet is high, whereas net gains of N occur when the ratio of N to DOM is low. This is consistent with the idea that microbial protein synthesis in the rumen is related to the availability of energy from the fermentation of DOM. As shown in the studies of Hogan & Weston (1970), with dried forage diets there is a curvilinear relationship between the ratio of duodenal NAN passage to dietary N intake and the ratio of N to DOM in the diet, and the results available for silage diets appear to be in accord with a similar relationship. However the range of N to DOM encountered in the silage diets so far examined is not as wide as that investigated by Hogan & Weston (1970), and there tends to be a direct linear relationship between duodenal passage of NAN and dietary N intake (fig. 5).

There is as yet only limited information on the degradability of dietary proteins and efficiency of microbial protein synthesis in animals given

silage diets, and few categorical statements may be made about these aspects of N digestion. The information available indicates that the degradability of silage crude protein varies with the additive used during ensilage and that, especially when formaldehyde is applied at a high rate, degradability is reduced (table 7). Initial results with formic acid silages (Thomas *et al.*, 1980c) indicated that the amount of silage protein passing undegraded to the duodenum was closely related to the proportion of true protein in the silage crude protein, and that there might be scope to alter protein degradability through ensilage techniques designed to influence the silage true protein N to non-protein N ratio. Investigation of this possibility has shown, however, that the rumen-undegradable protein in formic acid silages represents a resistant plant fraction that survives fermentation in the silo. Ensilage techniques using direct-cutting of the crop and high levels of formic acid addition at

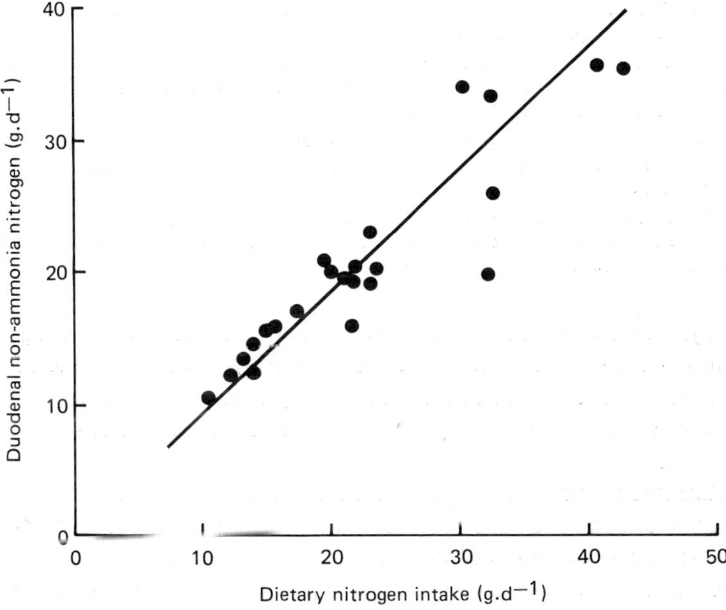


**Fig.5. The relationship between the duodenal passage of non-ammonia nitrogen (NAN) and the dietary intake in sheep given silage diets.** Values are those given in table 6. The straight line represents the values that would be obtained by assuming that duodenal NAN was, on average, 0.93 of nitrogen intake (see table 6)

**Table 7. Estimates of the rumen degradability of dietary crude protein in sheep and cattle given wilted or unwilted silages made with different additives (after Thomas *et al.*, 1980d)**

| Additive | Post-cutting treatment | Degradability | Source of data |
|---|---|---|---|
| None | Wilted | 0.69†‡ | Brett *et al.* (1979) |
| | Unwilted | 0.88§ | Beever *et al.* (1977) |
| | Wilted | 0.83‖ | Siddons *et al.* (1979) |
| Formic acid | Unwilted | 0.69†§ | Thomson *et al.* (1981) |
| | Wilted | 0.62¶ | Thomas *et al.* (1980c) |
| | Unwilted | 0.58¶ | Thomas *et al.* (1980c) |
| | Unwilted | 0.55¶ | Thomas *et al.* (1980c) |
| Formaldehyde or formaldehyde-formic acid | Unwilted | 0.52†§ | Thomson *et al.* (1981) |
| | Wilted | 0.33‖ | Siddons *et al.* (1979) |
| | Unwilted | 0.41† | Beever *et al.* (1977) |

† Experiments with dairy cows or growing cattle

‡ Calculated from the duodenal passage of nitrogenous constituents. Analytical techniques used to fractionate the digesta not given

§ Calculated from the duodenal passage of food protein as determined by the difference between total amino acid passage and the passage of microbial amino acids (estimated with $^{35}S$), and endogenous amino acids (assumed to be 0.11 of total amino acid passage)

‖ Calculated from the duodenal passage of food NAN as determined by the difference between total NAN and the passage of microbial nitrogen (estimated with $^{35}S$), and endogenous nitrogen (assumed to be 1.5 g. $d^{-1}$)

¶ Calculated from the duodenal passage of food NAN as determined by the difference between total NAN passage and the passage of bacterial nitrogen (estimated with $\alpha$-$\epsilon$-diaminopimelic acid), and protozoal and endogenous nitrogen (assumed to be 3 g. $d^{-1}$)

ensilage, which increase the proportion of true protein in the silage by limiting protein breakdown in the silo, merely result in an increased proportion of silage true protein being degraded in the rumen with no net benefit in the passage of protein to the small intestine (Chamberlain *et al.*, 1982).

Estimates of the efficiency of microbial protein synthesis in the rumen are summarized in table 8. The values indicate that for silage diets the efficiencies are low: compare the value of 30 g microbial N per kg organic matter given by the ARC (1980). This in part results from the fact that a portion of the DOM in silage is present as silage fermentation products and the energy yield from the rumen fermentation of these products is relatively poor. However, even taking account of this, the evidence suggests that in animals receiving diets containing a high proportion of silage the efficiency of microbial protein synthesis may be restricted by nutritional or other factors. There are substantial populations of proto-

**Table 8. The efficiency of microbial protein synthesis in the rumen in animals given silage diets**

| Diet | Efficiency of microbial protein synthesis (g N per kg organic matter apparently digested in the rumen) | Source of data |
|---|---|---|
| Wilted red clover silage made with formic acid plus barley (680:320 g. $kg^{-1}$) | 31.3 | Thomas *et al.* (1976a)‡§ |
| Wilted red clover silage made with formic acid plus barley-groundnut (680:320 g. $kg^{-1}$) | 31.2 | |
| Wilted red clover silage made with formic acid plus barley-urea (680:320 g. $kg^{-1}$) | 28.8 | |
| Unwilted ryegrass silage made without additive | 33.4 | Beever *et al.* (1977)‡ ‖ |
| Unwilted ryegrass silage made with formaldehyde additive | 13.1 | |
| Wilted ryegrass silage made with formic acid additive | 21.6 | Thomas *et al.* (1980c) |
| Wilted ryegrass silage made with formic acid additive plus barley (800:200 g. $kg^{-1}$) | 25.2 | |
| Unwilted ryegrass silage made with formic acid additive | 10.3 | |
| Unwilted ryegrass silage made with formic acid additive | 16.4 | Siddons *et al.* (1979)‡ ¶ |
| Wilted ryegrass silage made without additive | 21.7 | |
| Wilted ryegrass silage made with formic acid-formaldehyde additive | 27.2 | |
| Wilted ryegrass silage made with formic acid-formaldehyde additive plus urea supplement | 23.3 | |
| Wilted ryegrass silage made with formic acid-formaldehyde additive plus urea-soya bean meal supplement | 17.8 | |
| Wilted grass silage made without additive | 27.3 | Brett *et al.* (1979)†† |
| Wilted grass silage made without additive plus barley-dairy concentrate (460:540 g. $kg^{-1}$) | 27.7 | |
| Wilted grass silage made without additive plus barley-dairy concentrate-soya bean meal (450:550 g. $kg^{-1}$) | 34.8 | |
| Wilted grass silage made without additive | 25.3 | Armstrong (1980) ††¶ |
| Wilted grass silage made with formic acid-formaldehyde additive | 25.2 | |
| Wilted grass silage made with formic acid-formaldehyde additive plus soya bean meal | 31.0 | |
| **Mean for all diets** | 24.9† | |
| **s.e.** | 1.5 | |

† This value has been rounded to 25 for use in the equation on p.91
‡ Experiments with sheep
§ Calculated using $\alpha$ - $\epsilon$ -diaminopimelic acid as bacterial matter
‖ Calculated from the duodenal flow of microbial amino acids
¶ Calculated from the incorporation of $^{35}S$ into microbial protein
†† Experiments with cattle

zoa in the rumen in silage-fed animals and defaunation leads to a significant reduction in rumen ammonia concentration (Chamberlain *et al.*, 1980), but whether this is accompanied by an improvement in the efficiency of microbial growth has yet to be determined. In cows given diets of formic acid silage with concentrates, Brett *et al.* (1979) found that the efficiency of microbial synthesis was increased by the inclusion of soya bean meal in the diet. However, in sheep given similar silage, no response was found to a specific supplement of sulphur amino acids and, in sheep given formic acid-formaldehyde silage, Siddons *et al.* (1979) found no response to soya bean meal.

Because of the low rate of protein synthesis, the proportion of microbial protein in the total protein passing to the duodenum in animals given silage diets tends to be lower than is found typically in animals consuming diets based on hay. There are also some characteristic differences in the amino acid composition of the duodenal protein and, with silage diets, the protein contains lower proportions of arginine, lysine and especially methionine (table 9). Typically, as measured by the digestion of NAN, approximately 0.63 of the total amino acids passing to the duodenum are absorbed in the small intestine (table 6) although, for individual amino acids, the digestibility may vary quite widely from this figure (see Thomas *et al.*, 1980c).

**Table 9. The proportions of essential amino acids (g amino acid per kg determined amino acids) in the duodenal digesta of animals given diets containing silage or hay as forage**

| | Silage diets† | | Hay diets‡ | |
|---|---|---|---|---|
| | Mean | s.e. | Mean | s.e. |
| Threonine | 66 | 5 | 60 | 1 |
| Valine | 66 | 1 | 68 | 3 |
| Methionine | 17 | 1 | 32 | 1 |
| Isoleucine | 50 | 6 | 59 | 1 |
| Leucine | 80 | 11 | 79 | 2 |
| Phenylalanine | 51 | 5 | ND§ | |
| Histidine | 13 | 3 | ND | |
| Lysine | 68 | 4 | 85 | 1 |
| Arginine | 38 | 4 | 58 | 2 |

† Mean with s.e. for three diets of ryegrass silage alone, and one diet of ryegrass silage and barley (800:200 g. kg$^{-1}$) (from Thomas *et al.*, 1980c)

‡ Mean with s.e. for three diets consisting of hay and a 500:500 g. kg$^{-1}$ barley-maize mixture in proportions 1000:0, 700:300 and 400:600 g. kg$^{-1}$ (from Chamberlain & Thomas, 1979)

§ ND, not determined

*Protein rationing with silage diets*

The basis of the ARC (1980) system for protein rationing has been described earlier but it is appropriate to consider specifically its application to silage diets. Limitations on information prevent unequivocal conclusions as to the accuracy of the system but it is clear that errors in ration formulation might occur through difficulties in ascribing an appropriate degradability value to a particular sample of silage. Also, the available evidence indicates that the factors used within the rationing system as proposed may be inappropriate for silage diets. For example, in the new system the amount of microbial N absorbed in the small intestines and deposited in the tissues (TMN) is calculated according to the equation:

$$\begin{aligned} \text{TMN (g. d}^{-1}) &= \text{ME (MJ. d}^{-1}) \times \frac{1}{19.0 \times 0.82} \times 0.65 \times 30 \times 0.8 \times 0.7 \times 0.75 \\ &= 0.53\ \text{ME (MJ. d}^{-1}) \end{aligned}$$

where

ME = ME intake
19.0 = calorific value of DOM (MJ. $kg^{-1}$)
0.82 = ME/DE
0.65 = proportion of DOM apparently digested in the rumen
30 = rate of microbial protein synthesis as g N per kg OM digested in the rumen
0.8 = proportion of amino acid-N in microbial total N
0.7 = apparent digestibility of amino acid-N in the small intestine
0.75 = efficiency of utilization of absorbed amino acids.

However, on the basis of the information presented above, a corresponding but more appropriate equation for high-quality silage diets would appear to be:

$$\begin{aligned} \text{TMN (g. d}^{-1}) &= \text{ME (MJ. d}^{-1}) \times \frac{1}{19.3 \times 0.84} \times 0.7 \times 25 \times 0.8 \times 0.63 \times 0.75 \\ &= 0.41\ \text{ME (MJ. d}^{-1}). \end{aligned}$$

Clearly, further information is required to clarify which factors are most appropriate to particular dietary conditions.

## Silage and milk composition

In addition to contributing energy and protein to the cow's diet, particular dietary ingredients confer on the diet characteristics that influence the composition of the cow's milk. Milk fat is produced in the udder using fatty acids that are absorbed preformed from the blood and fatty acids synthesized *de novo* in the udder tissue. The preformed fatty acids mainly have a chain length of 16 or 18 carbon atoms and are derived largely from fatty acids of corresponding chain length absorbed from the intestine, whilst the fatty acids synthesized in the udder have chain lengths of 4 to 16 carbon atoms, and are derived mainly from acetate and butyrate absorbed from the rumen. Milk fat content and fatty acid composition is thus influenced by the amount and composition of fat in the diet, and by the pattern of fermentation in the rumen. Fermentation patterns high in propionic acid and low in acetic plus butyric acids, such as are obtained with diets containing a high proportion of concentrates and a low proportion of grass or hay, substantially reduce milk fat concentration (see Fogerty & Johnson, 1980; Storry, 1980; Storry *et al.*, 1980; Sutton, 1980).

Most of the proteins in milk are of specific types (caseins, $\alpha$-lactalbumin and $\beta$-lactoglobulin) synthesized in the udder, and sufficient free amino acids are absorbed from the blood to account for the total amount of amino acids necessary for the synthesis to take place. For some of the essential amino acids there is an almost quantitative incorporation of the absorbed amino acid into milk protein but oxidation, interconversion and synthesis of non-essential amino acids occurs within the udder tissue, and the relationship between amino acid supply and milk protein synthesis is not a simple one (see Thomas, 1980). Milk protein content is reduced by diets providing protein intakes substantially below the cow's requirements but protein content is not very sensitive to dietary protein supply and is not reduced by mild protein undernutrition, although milk yield is affected. In animals given hay and concentrate diets, milk protein content has been increased through additional allowances of concentrates, especially in the form of steamed cereal, such as flaked maize (see Rook, 1976; Thomas, 1980). This effect has been considered to result from an enhanced proportion of propionic acid in the rumen, since similar responses in protein content are obtained when propionic acid is given as an intraruminal infusion (Rook & Balch, 1961).

Until recently there has been little detailed information about the effects of silage diets on milk composition and there has been a particular lack of knowledge about diets containing a high proportion of silage, or silage of high digestibility. Some experiments on this subject have now been undertaken and special features of the relationship between diet composition and milk composition with silage diets have been identified. With diets adequate in energy and protein but containing a large proportion of silage (*ca*. 0.70 of the DM), with barley as the concentrate, the milk fat secreted has a high content of the fatty acids

**Table 10. Values for milk fat content and fatty acid composition of the milk fat (g. kg$^{-1}$) in Ayrshire cows given diets of silage and barley or silage and dairy concentrates. (Data are from unpublished experiments at the Hannah Research Institute)**

| | Diet (g. kg$^{-1}$) | | |
|---|---|---|---|
| | Silage:barley | Silage:dairy concentrate | |
| | 700:300 | 600:400 | 220:780 |
| Milk fat content | 51.4 | 39.3 | 35.0 |
| Fatty acid composition of the fat | | | |
| 6:0 | 26 | 25 | 26 |
| 8:0 | 16 | 17 | 17 |
| 10:0 | 32 | 33 | 38 |
| 12:0 | 40 | 40 | 44 |
| 14:0 | 136 | 139 | 136 |
| 16:0 | 493 | 409 | 339 |
| 18:0 | 79 | 100 | 111 |
| 18:1 | 164 | 216 | 257 |
| 18:2 | 10 | 12 | 19 |
| 18:3 | 5 | 12 | 14 |

formed by synthesis *de novo* in the udder and the milk is of a high fat content (see, for example, table 10). Under these circumstances an increase in the protein content of the concentrate leads to a reduction in fat content and to an increase in the proportion of $C_{18}$ acids in the fat (Kelly *et al.*, 1980). Moderately high proportions of $C_{18}$ acids are also observed in the milk fat of cows given diets of silage and concentrates typical of those found in commercial practice (table 10) and, as is the case for diets based on hay, milk fat content is reduced as the proportion of concentrate in the diet is increased. However, with silage diets, precipitous reductions in milk fat content with high concentrate feeding (often referred to as the 'low milk-fat syndrome') are not common, presumably because of the high stability of the fermentation in the rumen (see above).

With some high-digestibility silage diets, low milk protein contents (and thus low solids-not-fat contents) not associated with energy undernutrition have been observed (Murdoch & Rook, 1963; Thomas *et al.*, 1980b). The occurrence of this problem is unpredictable but, where it arises, it may be difficult to resolve through manipulation of the diet. As has been indicated earlier, the traditional remedial approach is to reduce the proportion of forage in the diet and to give additional flaked-maize concentrate to increase the proportion of propionic acid in the rumen. However, with silage diets, this type of dietary change will probably not produce the desired change in rumen fermentation pattern (see above). Furthermore, in contrast to the effects observed in animals given hay and concentrate diets, intraruminal infusions of propionic acid have not increased milk protein content in cows given silage (Chalmers *et al.*, 1980). This suggests that there may be fundamental differences between hay-based and silage-based diets in the relationships between rumen fermentation pattern and milk composition, although the basis of such differences is not yet understood.

## REFERENCES

AGRICULTURAL RESEARCH COUNCIL 1965 *The Nutrient Requirements of Farm Livestock. No. 2, Ruminants.* Agricultural Research Council, London

AGRICULTURAL RESEARCH COUNCIL 1976 *The Nutrient Requirements of Farm Livestock. No. 4, Composition of British Feedingstuffs.* Agricultural Research Council, London

AGRICULTURAL RESEARCH COUNCIL 1980 *The Nutrient Requirements of Ruminant Livestock.* Commonwealth Agricultural Bureaux, Slough

ALEXANDER, R.H. 1969 A laboratory homogenizer suitable for preparation of suspensions of wet fibrous materials such as silage. *Laboratory Practice* **18** 63-65

ARMSTRONG, D.G. 1980 Net efficiencies *in vivo* of microbial synthesis in ruminant livestock. *Proceedings, 3rd European Association of Animal Production Symposium on Protein Metabolism and Nutrition* (Eds H.J. Oslage & K. Rohr) Vol. II pp 400-413 Information Centre of Bundesforschungsanstalt für Landwirtschaft, Braunschweig

BAUMGARDT, B.R. 1970 Control of feed intake in the regulation of energy balance. In *Physiology of Digestion and Metabolism in the Ruminant* (Ed. A.T. Phillipson) pp 235-253 Oriel Press, Newcastle upon Tyne

BEEVER, D.E., THOMSON, D.J., PFEFFER, E. & ARMSTRONG, D.G. 1971 The effect of drying and ensiling grass on its digestion in sheep. Sites of energy and carbohydrate digestion. *British Journal of Nutrition* **26** 123-134

BEEVER, D.E., THOMSON, D.J., CAMMELL, S.B. & HARRISON, D.G. 1977 The digestion by sheep of silages made with and without the addition of formaldehyde. *Journal of Agricultural Science* **88** 61-70

BLAXTER, SIR KENNETH 1980 Feeds as sources of energy for ruminant animals. *Massey-Ferguson Papers* Massey-Ferguson, Coventry

BRETT, P.A., ALMOND, M., HARRISON, D.G., ROWLINSON, P., ROOKE, J. & ARMSTRONG, D.G. 1979 An attempted evaluation of the proposed ARC protein system with reference to the lactating cow. *Proceedings of the Nutrition Society* **38** 148A (Abstr.)

CAMPLING, R.C. 1966 The intake of hay and silage by cows. *Journal of the British Grassland Society* **21** 41-48

CASTLE, M.E., RETTER, W.C. & WATSON, J.N. 1979 Silage and milk production: comparisons between grass silage of three different chop lengths. *Grass and Forage Science* **34** 293-301

CHALMERS, J.S., THOMAS, P.C. & CHAMBERLAIN, D.G. 1980 The effect of intra-ruminal infusions of propionic acid on milk composition in cows given silage diets. *Proceedings of the Nutrition Society* **39** 27A (Abstr.)

CHAMBERLAIN, D.G. & THOMAS, P.C. 1979 Ruminal nitrogen metabolism and the passage of amino acids to the duodenum in sheep receiving diets containing hay and concentrates in various proportions. *Journal of the Science of Food and Agriculture* **30** 677-686

CHAMBERLAIN, D.G., THOMAS, P.C. & ANDERSON, FIONA J. 1980 Rumen fermentation pattern and protozoal counts in sheep given silage and silage and barley diets. *Proceedings of the Nutrition Society* **39** 29A (Abstr.)

CHAMBERLAIN, D.G., THOMAS, P.C. & WAIT, M.K. 1982 The rate of addition of formic acid to grass at ensilage and the subsequent digestion of the silage in the rumen and intestines of sheep. *Grass and Forage Science* **37** 159-164

CLANCY, M., WANGSNESS, P.J. & BAUMGARDT, B.R. 1977 Effect of silage extract on voluntary intake, rumen fluid constituents, and rumen motility. *Journal of Dairy Science* **60** 580-590

DESWYSEN, A., VANBELLE, M. & FOCANT, M. 1978 The effect of silage chop length on the voluntary intake and rumination behaviour of sheep. *Journal of the British Grassland Society* **33** 107-115

DEWAR, W.A. & McDONALD, P. 1961 Determination of dry matter in silage by distillation with toluene. *Journal of the Science of Food and Agriculture* **12** 790-795

DULPHY, J.-P. & DEMARQUILLY, C. 1973 [Effect of type of forage harvester and chopping fineness on the feeding value of silages.] *Annales de Zootechnie* **22** 199-217

EKERN, A. & SUNDSTØL, F. 1974 Energy utilization of hay and silages by sheep. In *Energy Metabolism of Farm Animals* (Eds K.H. Menke, H.-J. Lantzsch & J.R. Reichl) pp 221-224 Universität Hohenheim, Stuttgart

FARHAN, S.M.A. & THOMAS, P.C. 1978 The effect of partial neutralization of formic acid silages with sodium bicarbonate on their voluntary intake by cattle and sheep. *Journal of the British Grassland Society* **33** 151-158

FOGERTY, A.C. & JOHNSON, A.R. 1980 Influence of nutritional factors on the yield and content of milk fat: protected polyunsaturated fat in the diet. In *Factors Affecting the Yields and Contents of Milk Constituents of Commercial Importance. Bulletin, International Dairy Federation, Document* 125 pp 96-104

FORBES, J.M. 1980 Hormones and metabolites in the control of food intake. In *Digestive Physiology and Metabolism in Ruminants* (Eds Y. Ruckebusch & P. Thivend) pp 145-160 MTP Press, Lancaster

HOGAN, J.P. & WESTON, R.H. 1970 Quantitative aspects of microbial protein synthesis in the rumen. In *Physiology of Digestion and Metabolism in the Ruminant* (Ed. A.T. Phillipson) pp 474-485 Oriel Press, Newcastle upon Tyne

HUTCHINSON, K.J. & WILKINS, R.J. 1971 The voluntary intake of silage by sheep. II. The effects of acetate on silage intake. *Journal of Agricultural Science* **77** 539-543

JONES, D.I.H. & HAYWARD, MARGARET V. 1975 The effect of pepsin pretreatment of herbage on the prediction of dry matter digestibility from solubility in fungal cellulase solutions. *Journal of the Science of Food and Agriculture* **26** 711-718

KELLY, N.C. & THOMAS, P.C. 1978a The nutritive value of silages. Energy metabolism in sheep receiving diets of grass silage or grass silage and barley. *British Journal of Nutrition* **40** 205-219

KELLY, N.C. & THOMAS, P.C. 1978b Lactic acid metabolism in sheep receiving diets of grass silage. *Proceedings, 5th Silage Conference* (Ed. R.D. Harkess) pp 24-25 Hannah Research Institute, Ayr

KELLY, N.C., THOMAS, P.C. & CHAMBERLAIN, D.G. 1980 The effect of dietary inclusions of protein on milk secretion and nitrogen retention in cows given silage-barley diets. *Proceedings of the Nutrition Society* **39** 62A (Abstr.)

LANCASTER, R.J. & WILSON, R.K. 1975 Effect on intake of adding sodium bicarbonate to silage. *New Zealand Journal of Experimental Agriculture* **3** 203-206

McCULLOUGH, M.E. 1978 Silage – some general considerations. In *Fermentation of Silage – a Review* (Ed. M.E. McCullough) pp 1-26. National Feed Ingredients Association, West Des Moines, Ia

McDONALD, P. & EDWARDS, R.A. 1976 The influence of conservation methods on digestion and utilization of forages by ruminants. *Proceedings of the Nutrition Society* **35** 201-211

McDONALD, P., HENDERSON, A.R. & RALTON, I. 1973 Energy changes during ensilage. *Journal of the Science of Food and Agriculture* **24** 827-834

McLEOD, D.S., WILKINS, R.J. & RAYMOND, W.F. 1970 The voluntary intake by sheep and cattle of silage differing in free-acid content. *Journal of Agricultural Science* **75** 311-319

MARSH, R. 1979 The effects of wilting on fermentation in the silo and on the nutritive value of silage. *Grass and Forage Science* **34** 1-9

MINISTRY OF AGRICULTURE, FISHERIES AND FOOD, DEPARTMENT OF AGRICULTURE AND FISHERIES FOR SCOTLAND & DEPARTMENT OF AGRICULTURE FOR NORTHERN IRELAND 1975 Energy allowances and feeding systems for ruminants. *Technical Bulletin* 33 Her Majesty's Stationery Office, London

MORGAN, D.E. & BARBER, W.P. 1980 The adviser's approach to predicting the metabolizable energy value of feeds for ruminants. In *Recent Advances in Animal Nutrition-1979* (Eds W. Haresign & D. Lewis) pp 93-106 Butterworth, London

MORGAN, C.A., EDWARDS, R.A. & McDONALD, P. 1980 Effect of energy and nitrogen supplements on the metabolism and intake of silage. In *Forage Conservation in the 80's, Occasional Symposium* No. 11 (Ed C. Thomas) pp 363-368 British Grassland Society, Maidenhead

MORRISON, I.M. 1973 A note on the evaluation of the nutritive value of forage crops by the acetyl bromide technique. *Journal of the British Grassland Society* **28** 153-154

MURDOCH, J.C. & ROOK, J.A.F. 1963 A comparison of hay and silage for milk production. *Journal of Dairy Research* **30** 391-397

NIJKAMP, H.J. 1965 Some remarks about the determination of the heat of combustion and the carbon content of urine. In *Energy Metabolism* (Ed. K.L. Blaxter) pp 147-157 Academic Press, London

ROOK J.A.F. 1976 Nutritional influences on milk quality. In *Principles of Cattle Production* (Eds H. Swan & W.H. Broster) pp 221-236 Butterworth, London

ROOK, J.A.F. & BALCH, C.C. 1961 The effects of intraruminal infusions of acetic, propionic and butyric acids on the yield and composition of the milk of the cow. *British Journal of Nutrition* **15** 361-369

SIDDONS, R.C., EVANS, R.T. & BEEVER, D.E. 1979 The effect of formaldehyde treatment before ensiling on the digestion of wilted grass silage by sheep. *British Journal of Nutrition* **42** 535-545

SMITH, E.JILL & CLAPPERTON, J.L. 1981 The voluntary food intake of sheep when silage juice is infused into the rumen. *Proceedings of the Nutrition Society* **40** 22A (Abstr.)

SMITH, J.S., WAINMAN, F.W. & DEWEY, P.J.S. 1977 The energy value to sheep of three mixed grass silages. *Proceedings of the Nutrition Society* **36** 66A (Abstr.)

SMITH, E.JILL, CLAPPERTON, J.L. & ROOK, J.A.F. 1980 The rate of digestion of the dry matter of forage material conserved in different ways. *Proceedings of the Nutrition Society* **39** 68A (Abstr.)

STORRY, J.E. 1980 Influence of nutritional factors on the yield and content of milk fat: non-protected fat in the diet. In *Factors Affecting the Yields and Contents of Milk Constituents of Commercial Importance. Bulletin, International Dairy Federation, Document* 125 pp 88-95

STORRY, J.E., BRUMBY, P.E. & DUNKLEY, W.L. 1980 Influence of nutritional factors on the yield and content of milk fat: protected non-polyunsaturated fat in the diet. In *Factors Affecting the Yields and Contents of Milk Constituents of Commercial Importance. Bulletin, International Dairy Federation, Document* 125 pp 105-125

SUNDSTØL, F., EKERN, A., LINGVALL, P., LINDGREN, E. & BERTILSSON, J. 1980 Energy utilization in sheep fed grass silage and hay. In *Energy Metabolism* (Ed. L.E. Mount) pp 17-21 Butterworth, London

SUTTON, J.D. 1980 Influence of nutritional factors on the yield and content of milk fat: dietary components other than fat. In *Factors Affecting the Yields and Contents of Milk Constituents of Commercial Importance. Bulletin, International Dairy Federation, Document* 125 pp 126-134

SWIFT, R.W. 1957 The calorie value of TDN. *Journal of Animal Science* **16** 753-756

TAYLER, J.C. & WILKINS, R.J. 1976 Conserved forage – complement or competitor to concentrates. In *Principles of Cattle Production* (Eds H. Swan & W.H. Broster) pp 343-364 Butterworth, London

TERRY, R.A. & OSBOURN, D.F. 1980 Determination and prediction of the digestible energy in silages. In *Forage Conservation in the 80's, Occasional Symposium* No. 11 (Ed. C. Thomas) pp 315-318 British Grassland Society, Maidenhead

THOMAS, P.C. 1973 Microbial protein synthesis. *Proceedings of the Nutrition Society* **32** 85-91

THOMAS, P.C. 1977 Ruminal fermentation and the flow of nitrogen compounds to the duodenum. *Proceedings, 2nd International Symposium on Protein Metabolism and Nutrition* (ed. S. Tamminga) pp 47-50 Centre for Agricultural Publishing and Documentation, Wageningen

THOMAS, P.C. 1980 Influence of nutrition on the yield and content of protein in milk: dietary protein and energy supply. In *Factors Affecting the Yields and Contents of Milk Constituents of Commercial Importance. Bulletin, International Dairy Federation, Document* 125 pp 142-151

THOMAS, P.C. & ROOK, J.A.F. 1977 Manipulation of rumen fermentation. In *Recent Advances in Animal Nutrition-1977* (Eds W. Haresign & D. Lewis) pp 83-109 Butterworth, London

THOMAS, C. & WILKINSON, J.M. 1975 The utilization of maize silage for intensive beef production. 3. Nitrogen and acidity as factors affecting the nutritive value of ensiled maize. *Journal of Agricultural Science* **85** 255-261

THOMAS, P.C., CHAMBERLAIN, D.G. & ALWASH, A.H. 1976a Digestion in the stomach and intestines of sheep receiving diets of red clover silage with various supplements. *Journal of the British Grassland Society* **31** 123-128

THOMAS, P.C., KELLY, N.C. & WAIT, M.K. 1976b The effect of physical form of a silage on its voluntary consumption and digestibility by sheep. *Journal of the British Grassland Society* **31** 19-22

THOMAS, C., GILL, M. & AUSTIN, A.R. 1980a The effect of supplements of fishmeal and lactic acid on voluntary intake of silage by calves. *Grass and Forage Science* **35** 275-279

THOMAS, P.C., CHALMERS, J.S., CHAMBERLAIN, D.G. & BELIBASAKIS, N. 1980b The effect of diet on the content and composition of crude protein in milk. *Proceedings, 3rd European Association of Animal Production Symposium on Protein Metabolism and Nutrition* (Eds H.J. Oslage & K. Rohr) Vol. II pp 522-528 Information Centre of Bundesforschungsanstalt für Landwirtschaft, Braunschweig

THOMAS, P.C., CHAMBERLAIN, D.G., KELLY, N.C. & WAIT, M.K. 1980c The nutritive value of silages. Digestion of nitrogenous constituents in sheep receiving diets of grass silage and grass silage and barley. *British Journal of Nutrition* **43** 469-479

THOMAS, P.C., KELLY, N.C. & CHAMBERLAIN, D.G. 1980d Silage. *Proceedings of the Nutrition Society* **39** 257-264

THOMAS, P.C., KELLY, N.C., CHAMBERLAIN, D.G. & CHALMERS, J.S. 1980e Some aspects of energy and protein utilization in ruminants given silage diets. In *Energy Metabolism* (Ed. L.E. Mount) pp 357-361 Butterworth, London

THOMAS, P.C., KELLY, N.C., CHAMBERLAIN, D.G. & WAIT, M.K. 1980f The nutritive value of silages. Digestion of organic matter, gross energy and carbohydrate constituents in the rumen and intestines of sheep receiving diets of grass silage or grass silage and barley. *British Journal of Nutrition* **43** 481-489

THOMSON, D.J., BEEVER, D.E., LONSDALE, C.R., HAINES, M.J., CAMMELL, S.B. & AUSTIN, A.R. 1981 The digestion by cattle of grass silage made with formic acid and formic acid-formaldehyde. *British Journal of Nutrition* **46** 193-207

TILLEY, J.M.A. & TERRY, R.A. 1963 A two-stage technique for the *in vitro* digestion of forage crops. *Journal of the British Grassland Society* **18** 104-111

UNSWORTH, E.F. & STEVENSON, M.H. 1978 Preliminary studies on the digestion of silages by surgically modified sheep. *Record of Agricultural Research* **26** 63-68

VAN ES, A.J.H. & NIJKAMP, H.J. 1969 Energy, carbon and nitrogen balance experiments with lactating cows. In *Energy Metabolism of Farm Animals* (Eds K.L. Blaxter, J. Kielanowski & Greta Thorbek) pp 209-212 Oriel Press, Newcastle upon Tyne

VETTER, R.L. & VON GLAN, K.N. 1978 Abnormal silages and silage related disease problems. In *Fermentation of Silage – a Review* (Ed. M.E. McCullough) pp 281-332 National Feed Ingredients Association, West Des Moines, Ia

WAINMAN, F.W., DEWEY, P.J.S. & BOYNE, A.W. 1979 *Feedingstuffs Evaluation*

*Unit, Second Report 1978* Department of Agriculture and Fisheries for Scotland, Edinburgh

WATSON, S.J. & NASH, M.J. 1960 *The Conservation of Grass and Forage Crops* 2nd ed. pp 691-695 Oliver and Boyd, Edinburgh

WEBSTER, A.J.F. 1980 Energy costs of digestion and metabolism in the gut. In *Digestive Physiology and Metabolism in Ruminants.* (Eds Y. Ruckebusch & P. Thivend) pp 469-484 MTP Press, Lancaster

WILKINS, R.J. 1974 The nutritive value of silages. *University of Nottingham Nutrition Conference for Feed Manufacturers:* 8 (Eds H. Swan & D. Lewis) pp 167-189 Butterworth, London

WILKINS, R.J. 1978 Ensiled forages and their utilization by ruminants. *Proceedings, 3rd World Conference on Animal Feeding, Madrid* Vol. 7 pp 403-412

WILKINS, R.J., HUTCHINSON, K.J., WILSON, R.F. & HARRIS, C.E. 1971 The voluntary intake of silage by sheep. I. Interrelationships between silage composition and intake. *Journal of Agricultural Science* **77** 531-537

WILKINS, R.J., FENLON, J.S., COOK, J.E. & WILSON, R.F. 1978 A further analysis of relationships between silage composition and voluntary intake by sheep. *Proceedings, 5th Silage Conference,* (Ed. R.D. Harkess) pp 34-35 Hannah Research Institute, Ayr

WOODMAN, H.E. 1948 Rations for livestock. *Bulletin, Ministry of Agriculture and Fisheries* No. 48 His Majesty's Stationery Office, London

ZIMMER, E. 1966 [A revision of Flieg's silage-evaluation key.] *Wirtschaftseigene Futter* **12** 299-303

# PRODUCTION AND USE OF HIGH-QUALITY SILAGE

# Chapter 5

# Making high-quality silage

**M.E. Castle**

*Hannah Research Institute, Ayr KA6 5HL*

The first two grass silages with a high concentration of digestible organic matter in the dry matter (DOMD) made at the Hannah Research Institute (HRI) were produced in 1970 in 30-t vacuum-pack silos (Castle & Watson, 1971). The silages had DOMD concentrations *in vitro* of 719 and 687 g. $kg^{-1}$, and gave the highest daily yields of milk recorded up to that time in the series of silage-feeding experiments started in 1967. In view of the success of these two silages, a determined effort was then made to produce, regularly each year, silages of high DOMD concentration on a commercial scale, i.e. in silos containing 100-300 t silage. Since then, 15 silages with a mean DOMD concentration of 694 g. $kg^{-1}$ have been produced and offered to dairy cows in a continuing series of feeding experiments. From the results of these and other experiments elsewhere it has been possible to formulate a series of recommendations for making silage with a high DOMD concentration, high intake characteristics and relatively low losses in the silo. Success cannot always be achieved, however, and in 1976 silage quality was lower than average, with a DOMD concentration of 646 g. $kg^{-1}$. This was attributed to the extremely wet and sunless conditions at the time of silage making. However, on the credit side, silage of high quality has now been produced in 9 years out of 10. The techniques for making high-quality silage based on research work and experience at the HRI will be discussed in this chapter.

The successful conservation of high-quality silage depends on careful attention to detail at every stage in the making process and, in addition, the application of the basic principles of silage making. Briefly, the ideal silage process aims first at anaerobic conditions, i.e. the exclusion of air at all times, and secondly, sufficient acid in the silage to restrict the activities of undesirable bacteria. Thus, almost every technique in making good silage is aimed either directly or indirectly at the target of acidity control under anaerobic conditions. These basic principles of silage making are always the same, irrespective of the crop, and attention must be paid to them throughout the entire process of growing and harvesting the crop, and filling and sealing the silo. Lack of attention to detail in applying these simple principles, including the correct design and the construction of the silo, may result in both increased conservation losses and a product with a much reduced feeding value. Details of the biochemical changes in the cut herbage and the fermentation in the silo have been given already (page 21), but it is stressed that the rapid initial development of acidity is vital to stop the undesirable micro-organisms growing. A poor silage preservation resulting in high levels of ammonia is associated with reduced digestibility, as illustrated in table 1 (Flynn, 1979). Silage making is a controlled process of fermentation, and the following suggestions should ensure that the silage is made efficiently with the minimum loss of feeding value.

**Table 1. The association between measures of silage preservation and digestibility from 14 paired comparisons (Flynn, 1979)**

| | Silage preservation | |
|---|---|---|
| | **Good** | **Bad** |
| Dry matter (DM) (g. kg$^{-1}$) | 220 | 212 |
| pH | 4.1 | 4.9 |
| Ammonia-nitrogen as proportion of total N | 0.103 | 0.245 |
| Butyric acid, proportion in DM | 0.003 | 0.023 |
| Digestibility of DM | 0.679 | 0.626 |

## THE SWARD AND ITS PREPARATION

As indicated in Chapter 3, the preferred choice of grass for silage making is perennial ryegrass (*Lolium perenne*). This grass is persistent, responds exceedingly well to applications of fertilizer-nitrogen and, equally important, has usually a high content of soluble sugars. There is also some evidence that the DOMD content of herbage and silage increases as the proportion of perennial ryegrass in the sward increases (Grant, 1978). In the last 6 years virtually all the grass silage made at the HRI has been from swards of pure perennial ryegrass, and this herbage has been the basis of the high-quality silage made in this period.

The sugar content of grass is reduced as the level of the N application is increased (page 54), but a rate of approximately 100 kg fertilizer-N per ha will generally result in a sugar content in perennial ryegrass that ensures a satisfactory fermentation. Since 1968 at the HRI, the mean rate of fertilizer-N application in spring has been 98 kg. ha$^{-1}$ for silage cuts in May and June, and excellent silages with good fermentation patterns have been produced. In the last 3 years the application rate has increased slightly to 109 kg N per ha, with equally satisfactory silages being made.

Many commercial silages are made from grass that may receive up to 120 kg N per ha in spring. For example, in the survey of Grant (1978) more than 0.50 of fields in 1975 and 0.33 in 1976 received more than 150 kg N per ha before cutting. Under experimental conditions, an excellent silage was made from herbage that had received 211 kg fertilizer-N per ha 5-6 weeks previously (Castle & Watson, 1969). This silage had a pH of 4.3, and contained 214 g of crude protein per kg dry matter (DM) with 452 g non-protein N per kg total N. Clearly, as with this silage, a high level of fertilizer-N may be applied to the grass and an excellent silage may be produced if careful attention is paid to all the other aspects of the silage-making process.

In assessing the rate of fertilizer-N to apply, in particular to the first cut of silage, recognition should be made of the nutrients in any slurry applied to the sward in the late winter and spring. At the HRI the slurry is applied when the herbage is short, as this avoids any contamination of the herbage leaf that might eventually be cut and ensiled. Contamination may also occur if farm-yard manure is applied late to fields that are to be cut for silage and it is always preferable to apply slurry rather than farm-yard manure on grassland for silage making.

The height of cutting has been discussed in Chapter 3 (page 50) and it was stressed that soil contamination due to close cutting must be avoided. It is thus helpful if grass fields are as level as possible. On the HRI farm the fields for reseeding are ploughed in one direction to avoid unnecessary ridges and hollows, and an effort is made at levelling. Mole hills are harrowed before herbage growth starts, and ruts are filled and rolled. Short stubbles increase the risk of soil contamination through scalping of the sward, and in dry weather some types of flail forage harvester may act like a vacuum cleaner and incorporate soil into the cut herbage. In practice, therefore, it is wise to leave a stubble of approximately 50 mm, although field topography, machinery and local experience will all dictate the optimum cutting height. Picking up, particularly by a double-chop forage harvester, is easier with a long stubble and the rate of wilting will be increased also. High concentrations of ash may indicate excessive soil contamination but an average value of 84 g ash per kg in the DM has been recorded in the 11 grass silages of high quality made at the HRI in the last 10 years. A mean value of 90 g ash per kg was found by Grant (1978) in silages from 46 farms in 1975 and 1976 where quality silage was being made. It would seem that values in excess of approximately 90-100 g ash per kg indicate contamination with soil but the best indication is a silica determination. Grass silage cannot be regarded as a major source of minerals, in particular phosphorus, for lactating cows (Stevenson & Unsworth, 1978), and a high content of ash indicates a contaminated silage with the possibility of poor fermentation characteristics, waste and low intake.

The advantages of grazing a silage sward in the winter months were discussed in Chapter 3 (page 51). On balance, it would appear worthwhile to remove old herbage from silage swards by grazing with sheep in the winter, and this practice is normally adopted at the HRI if high-quality silage is the ultimate aim. This grazing is not continued late into the spring, i.e. after the end of March, so that the yields of herbage for silage are not seriously lowered. Grazing with dairy cows in the spring may reduce later silage yields (Grant, 1978) and this practice is avoided.

Finally, in spring it is vital to remove any debris from the silage fields and to roll them when the soil is still soft. Metal objects and large stones may seriously damage silage machinery and should be completely removed; small objects and stones can be rolled out of sight.

## CUTTING AND WILTING

Herbage for silage making may be either cut directly with the forage harvester, usually a flail-type machine, and blown into a trailer, or cut with a mowing machine, dried partially in the field, i.e. wilted, and then picked up by another machine prior to ensiling. The effects of wilting have been fully reviewed by Marsh (1979), who concluded that this process increased fermentation quality and animal intake, and was generally to be recommended. Field losses of DM associated with wilting in moderate to good conditions average approximately 20-30 g. $kg^{-1}$. $d^{-1}$ plus a constant amount of 20-30 g. $kg^{-1}$ if the crop has been lacerated. However, against these field losses may be set the reduction in storage losses as effluent. Thus, on balance, wilting is not necessarily disadvantageous in terms of total DM losses. In 40 comparisons of wilted and unwilted silage made from the same sward, an average decrease of 15 g. $kg^{-1}$ in the digestibility of the DM was found (Marsh, 1979). However, if the grass is on the ground for over 5 d it can readily lose 50-100 g. $kg^{-1}$ DM digestibility (Flynn,1979). Silage DM content is not itself a major factor determining silage feeding value, and wilting should be adopted mainly to avoid effluent production and to allow the storage of more DM in a fixed volume of silo. The effect of wilting on animal production is discussed in Chapter 6.

If the crop is to be wilted, cutting is now usually done with a rotary mower, i.e. a drum or a disc mower. Reciprocating-blade mowers are outdated because of their low work rate per h and high maintenance costs, and flail mowers increase DM losses in the field. The output of rotary and flail mowers is similar, and depends largely on the cutting width. The wider the swathe, the larger the area that can be cut per h, but more tractor power is required and more capital has to be invested.

Mower-conditioners also increase the demand for power, but in good weather these machines can certainly increase the wilting rate of the cut herbage. Unfortunately, in wet weather, herbage that has been conditioned may absorb moisture more rapidly than normally-cut herbage. Flail mowers appear to be the most efficient machines for rapidly reducing the moisture content of herbage, but these machines cause the largest reduction in the DOMD content of the cut herbage and give the largest losses in the field. The rate of wilting appears to be least with herbage cut with a drum mower regardless of the length of the period of wilting.

In the survey of Grant (1978) a wilting period of up to 18 h increased the DM concentration of the cut herbage by an average of 36 g. $kg^{-1}$, 24-36 h by 70 g. $kg^{-1}$ and over 48 h by 95 g. $kg^{-1}$. As expected, the increase in DM concentration was greatest in the driest year, and with herbage with an initially higher DM content. The rate of wilting depends on the micro-climate within the cut swathe. If the humidity is high within the swathe an equilibrium develops between the air and the grass, and no further drying of the herbage can occur. Sun and wind will rapidly alter this equilibrium and then drying is speeded up. Drying is normally more rapid on a long than on a short stubble. Where possible, the grass should be cut when the plant is dry, i.e. avoiding cutting in the early morning and late at night, when herbage may be damp. For this reason, and also to ensure the maximum concentration of sugar in the crop, it is theoretically advantageous to cut herbage in the driest of the daylight hours when the sun is shining. In practice this is not always possible, but if the aim is a good fermentation in the silage, it is well worth while to cut the crop under as good conditions as possible, even if this means a slight delay in the date of cutting.

In recent years all the herbage for high-quality silage made at the HRI has been cut with a disc mower, and satisfactory rates of wilting have been achieved. The mean length of the wilting period for the last 11 high-quality silages has been 34 h and this has resulted in herbage as ensiled with a mean DM concentration of 261 g. $kg^{-1}$. The lowest DM content occurred with grass cut in late May 1976 when 35 mm rain fell in the 'wilting' period. The resultant silage had a DOMD concentration of only 646 g. $kg^{-1}$ — the lowest of any recently made silage. The wet herbage was not the sole reason for the low DOMD value, but this particular silage produced large volumes of effluent as a result of ensiling herbage with a mean DM concentration of only 168 g. $kg^{-1}$. A target value of approximately 250 g DM per kg in the herbage at the time of ensiling is thus the aim.

Depending on the weather, the weight of the herbage in the swathe and the type of forage harvester, two swathes may sometimes be placed together. This operation involves extra labour and more capital equipment but will ensure a larger flow of grass to the forage harvester. Thus, ultimately, more grass per h could be harvested, and the precision-chop harvester would be working harder and more efficiently. The major drawback to putting two swathes into one is the great risk of incorporating soil into the herbage and hence increasing the chances of a poor fermentation. Care has to be taken during this operation to avoid running the tractor and machinery wheels over the cut swathe. Some machinery for rowing the crop is driven by contact with the soil and may also incorporate stones into the swathe; this should be avoided. The best machines are powered directly by the tractor, and only touch the grass and not the soil. Undue and excessive agitation of a swathe should be avoided. Primarily to avoid soil contamination, a stubble height of approximately 50 mm is ideal, although a few specialized silage makers, who aim for a high DM silage, cut the crop to leave an initial stubble height of 100 mm as an aid to drying. The swathe is turned and then picked up by a flail harvester, which removes the dry swathe and some of the previously uncut herbage. This technique may not be applicable universally, but it is a vital part of one successful system of quality-silage making.

Wilting concentrates the sugars in the cut herbage and the effect of different levels of water-soluble carbohydrates (sugars) on silage fermentation is shown in table 2. The DM proportion of both crops was low but this example indicates how a high sugar concentration may result in a high lactic acid concentration, a low pH, low ammonia and low organic acid concentration, i.e. a better silage fermentation than that from a crop low in sugars. Wilting may thus reduce the organic acids and the $NH_3$ in silage. A level of 25-30 g sugars per kg in the crop as ensiled is widely advocated but this measurement can rarely be made in commercial farming practice. Schemes for sampling and analysing herbage for water-soluble carbohydrates have been tried but have not been adopted to any significant extent. An alternative is to grow a ryegrass, to cut the crop in sunny weather if possible, and to wilt for 24 h.

In order to cut the herbage at a predetermined DOMD value it is possible to take small samples from the sward at regular intervals of time

**Table 2.** The composition of silages made from crops with high and low concentrations of water-soluble carbohydrate (from Wilson & Wilkins, 1973)

| Crop | Composition of crop | | Composition of silage | | | | |
|---|---|---|---|---|---|---|---|
| | | | | Acid (proportion of DM)(g. $kg^{-1}$) | | | |
| | DM (g. $kg^{-1}$ | Water-soluble carbohydrate in DM (g. $kg^{-1}$) | pH | Lactic | Acetic | Butyric | $NH_3$-N as proportion of total N (g. $kg^{-1}$) |
| Ryegrass | 163 | 164 | 3.8 | 154 | 34 | 0 | 85 |
| Lucerne | 163 | 45 | 6.4 | 21 | 65 | 29 | 244 |

and to determine the DOMD concentration by a rapid technique, e.g. Morrison (1972). This can be done under specialized conditions, such as at a research institute, but not generally in farming practice. A far more rapid method (Walters, 1976) for assessing the DOMD concentration of grass in the field is based on visual appraisal of the stage of growth of the grass and the proportion of leaf. The criteria used in the method are summarized in table 3. Success clearly depends on good sampling in the field, but the method gives a rapid and simple indication of DOMD, allowing the date for cutting the herbage to be decided.

The date of cutting (see Chapter 3) has a large effect on the DOMD concentration, and in order to obtain silages of high quality the grass at the HRI is normally cut in late May and early June for the first harvest. Swards of S24 perennial ryegrass have, in the last few years, been cut between 16 and 31 May with a mean date of 23 May, when there were no flower heads or just a few starting to emerge. The later S23 perennial ryegrass swards were cut between 26 May and 3 June with a mean date of 28 May, when no flower heads were visible, and the crop was dense and leafy.

The average yield for 11 recent silages made in the spring was 4520 kg DM per ha. This is equivalent to approximately 15 t of silage with a DM content of 250 g. kg$^{-1}$, after allowing for normal losses in the silo, and is sufficient to feed two cows for the winter-feeding period. The yields from S24, S23 and mixed swards were 4700, 4430 and 4270 kg DM per ha respectively. The date of the first cut determines not only the yield and quality of the first crop, but also the proportion of the total annual yield contributed by that crop. In general, as the primary cutting date is delayed, the first crop assumes an increasing proportion of the total yield and should not therefore be delayed unduly. The exact date of

**Table 3. Method for estimating the DOMD value (g. kg$^{-1}$) of herbage from the stage of growth and the proportion of leaf (Walters, 1976)**

| Weight of leaf as a proportion of total crop (g. kg$^{-1}$) | No flower heads emerging | Flower heads just emerging | Flower heads 0.75 emerged | Heads emerged and free from top leaf | Heads emerged and flower stalks elongating | Heads emerged and anthers visible |
|---|---|---|---|---|---|---|
| 100 | – | – | – | – | 570 | 550 |
| 200 | – | – | 630 | 610 | 600 | 580 |
| 300 | 690 | 670 | 660 | 640 | 630 | 610 |
| 400 | 720 | 700 | 690 | 670 | – | – |
| 500 | 750 | 730 | – | – | – | – |

cutting cannot be specified for every situation, but the aim must be to produce sufficient silage of the correct quality for the particular animal-production system adopted on the farm.

## USE OF ADDITIVES

An acceptable silage with satisfactory fermentation characteristics can be made without the use of additives, especially if ryegrass is used, and the crop is wilted, chopped and well consolidated. However, extensive experimental evidence and much practical experience clearly indicate the considerable benefits that additives can confer. However, no additive, however efficient, can rectify a basic fault in silage-making technique, such as cutting too late or poor sealing. Silage additives control, and thus improve, the fermentation in the silo, reduce waste and make a more acceptable product for livestock.

There are three main types of additives: those that encourage fermentation by supplying extra carbohydrates to be converted to acids; those, such as acids, that provide conditions that favour the acid-producing lactobacilli; and those that act as sterilants and either prevent or severely restrict silage fermentation.

Molasses, which contains approximately 500 g sugars per kg, was used widely as a silage additive until acids came on the market. However, with the present interest in safe additives and the development of molasses applicators, molasses is now being used increasingly. The rate of application varies from 10 to 20 l molasses per t, with the highest rate being applied to young, wet and leafy crops. The molasses must be thoroughly incorporated with the crop.

Acids, in particular formic acid, have proved to be highly effective in making first-class silage with good fermentation characteristics. Their use has been facilitated particularly by the development of applicators that can mix small and predetermined amounts of additive into the grass at the time of harvesting. Formic acid, at an average rate of 2.25 $l.t^{-1}$, has been used on virtually all the high-digestibility silage made at the HRI in recent years and has been most effective. Thirteen silages with DM concentrations below 300 $g.kg^{-1}$ in the last few years had a mean pH of 3.93, with only 93 $g.kg^{-1}$ total N in the form of $NH_3$. This indicates a satisfactory fermentation and the silages have had excellent intake characteristics (Chapter 6). Briefly, the formic acid lowers rapidly the pH of the herbage, respiration is restricted and conditions favour the lactobacilli. In addition, the temperature of the silage is reduced, as shown by experiments in which identical herbage was ensiled either with or without 2.2 l formic acid per t. Five acid-treated silages had a pH of 4.1 and a DM concentration of 225 g. kg $^{-1}$, whereas five control silages had a pH of 4.8 and a DM concentration of 216 g. kg $^{-1}$ (Castle & Watson, 1970a & b & 1973a). In addition, the acid-treated silages had higher DOMD values. Formic acid also restricts clostridial activity, and thus the concentrations of $NH_3$ and fermentation acids are lower than in untreated silages. Degradation of protein to amino acids is reduced slightly by formic acid treatment (Wilkinson *et al.*, 1976) and there is evidence that N retention is increased following the addition of formic acid to direct-cut silage (Waldo *et al.*, 1971).

Formalin, which contains 370-400 g formaldehyde per l, has some value as an additive, but it is not an ideal material on its own (Wilkins *et al.*, 1974). It will restrict protein degradation and the fermentation of the soluble carbohydrates in the silo but the level of application is critical. At low levels of application, approximately 2.2 l. t $^{-1}$, it can promote a clostridial fermentation, whereas at a high level, 4.5 l. t $^{-1}$ and above, it can decrease the contents of $NH_3$ and fermentation acids but may lower the availability of the silage protein to the animal. Formalin is also unpleasant to handle, especially in a confined space. One method of overcoming the problems with formalin, in particular the risk of a clostridial fermentation, is to mix it with a relatively small quantity of acid. This has been done with the commercial product 'Sylade' (Imperial Chemical Industries Ltd), which contains formalin and sulphuric acid, with recommended application rates of 2.5-5.0 l. t $^{-1}$. In a comparative feeding trial, silages treated with formic acid ('Add-F', BP Nutrition Ltd) at 2.0 l. t $^{-1}$, and a formalin/sulphuric acid mixture ('Sylade') at rates of 2.0 and 4.4 l. t $^{-1}$ (Castle *et al.*, 1977), had an average DOMD value of 709 g. kg $^{-1}$, pH values less than 4.0 and all contained less than 100 g $NH_3$-N per kg total N. The formalin additive did not appear to have any specific beneficial effect on the protein in the silage when compared with the formic acid. It has been suggested that a safe and effective working range for the application of formaldehyde as an additive for grasses is 30 to 50 g. kg $^{-1}$ crude protein (CP) (Wilkinson *et al.*, 1976). This is equivalent to 2.0 and 3.4 l formalin per t fresh grass with a DM concentration of 170 g. kg $^{-1}$, and 140 g CP per kg DM. At this level of addition, a clostridial fermentation is likely to occur and in practice an acid is also required.

New formulations of commercial additives appear regularly on the market, and new chemicals are being used as silage additives, e.g. acrylic acid (Wilson *et al.*, 1979). The ultimate choice of a particular additive should, however, be decided largely on the basis of published experiments in which animal production has been the final arbiter. Then, finally, the extra cost of the additive must be considered in relation to the value of the increased production of the livestock due to feeding the improved silage. It is not easy to measure the effect of an additive unless there is a control silage without an additive but most experimental work has shown the economic advantages of using a suitable additive. It is noteworthy that in the survey of quality-silage making (Grant, 1978), additives were used on 0.92 of crops in 1975 and 0.88 in 1976.

The success with additives has been achieved partly by the development of commercial applicators, which mix an accurate amount of additive uniformly with the crop. A simple, trouble-free applicator is a vital part of the silage-making equipment. For effective mixing, the additive should enter the forage harvester at the point of maximum air-turbulence, and this site may be indicated by the manufacturer of the equipment. For most liquid additives, which are applied at rates of 2.2-5.0 l. $t^{-1}$, the applicator consists of the inverted container of additive, a simple on-off mechanism and a variable-sized nozzle for flow control. These applicators require little maintenance and will last for many years. Slightly more sophisticated applicators have been designed and used at the HRI (Czerkawski *et al.*, 1977) using a peristaltic pump. This equipment enables a large container of additive to be retained in a safe and upright position, and gives a more accurate rate of application. Applicators, regardless of their design, should be simple to operate, have a minimum of moving parts, and should be capable of the accurate and uniform application of the additive.

Microbial cultures in the form of dry powders containing lactic acid bacteria are now available but have not been used at the HRI. The limited experimental data do not indicate that the present commercial cultures show any real benefits.

## CHOPPING AND LACERATION

In the early days of silage making, the grass crop was ensiled mainly in the long state with no attempt to lacerate and chop the material. As time has passed, the crop has been subjected to an increasing degree of mechanical treatment, and at the present time it is possible to find herbage that is precision-chopped to a length of approximately 10 mm before it is ensiled. This development over the last 30 years has had numerous beneficial effects on the resultant silage, although the costs of the silage-making machinery have increased.

Chopping herbage prior to ensiling has three main effects on the silage fermentation process: the crop is inoculated with bacteria, plant juices are liberated and the oxygen content of the crop is reduced. Microbiological studies have shown that lactic acid bacteria are scarce on fresh growing herbage, and only occur on damaged herbage and partially decayed material (Stirling & Whittenbury, 1963). However, when grass is passed through machinery such as a forage harvester, grass juice collects on the metal parts and it is an excellent medium for the development of the lactic acid bacteria. As a result, in one study, the number of bacteria increased from $10^5$ per kg fresh grass to approximately $5 \times 10^8$ per kg after forage harvesting (McDonald, 1976). The release of intracellular plant juices under anaerobic conditions is important for the onset of fermentation and thus physical damage to the crop is advantageous. The number of bacteria in minced grass is higher than in unchopped grass and hence the pH of the minced grass will be lower than that of the unchopped material. This rapid development of a low pH is important in making high-quality silage and, in turn, influences the type of organisms that grow. The more rapidly the plant juices are liberated, the more actively the lactic acid bacteria compete with other species. If a crop is finely chopped, the consolidation in the silo will also be improved, less oxygen will be present and losses from oxidation should be reduced. An anaerobic type of bacterial population is required, and short chopping and consolidation will aid this.

Physical treatment of forage during harvesting will thus generally result in silage with a higher lactic acid content and a lower pH, butyric acid and $NH_3$-N concentration than found with unchopped or coarse-chopped material, although it is difficult to be precise about the quantitative significance of this effect under farm conditions. Marsh (1978) pointed out that different results may be obtained in laboratory-scale silage experiments and in field-scale trials with large silos, and it has been suggested that, at a farm-scale, compaction of herbage in the silo may reduce the differential effects of physical treatment on silage quality. Nonetheless, in an experiment at the HRI using silos holding 40-50 t, long-chopped silage (72 mm) had a higher pH, a lower lactic acid and a higher butyric acid concentration than either medium- (17 mm) or short-chopped (9 mm) silages made from the same sward.

The effects of chopping, laceration and type of forage harvester on losses in the silo are somewhat conflicting (Marsh, 1978) but are explained possibly by the way in which the silo is sealed and hence air is excluded. If the silo is well consolidated and has an excellent top-seal to avoid air movement, different chop lengths are not likely to greatly effect total losses from the silo. If, however, the sealing is poor and consolidation is not good, a short-chopped crop would be better than a long unchopped or a lightly-lacerated crop. Normal physical treatment of herbage for silage making does not appear to have any large and consistent effect on digestibility, but the results tend to be confounded with the level of silage intake. In the recent experiment at the HRI with silages of 9, 17 and 72 mm the DOMD values *in vitro* were 659, 652 and 644 g. $kg^{-1}$ respectively, but further animal studies at several levels of controlled intake are necessary before the effect of physical treatment may be stated with certainty. It may be concluded that chopping and laceration of herbage is clearly beneficial to silage fermentation, and to the biochemical processes within the silo, and for these reasons is recommended.

It is worth-while to chop as short as possible without seriously limiting the rate at which the herbage is collected from the field. At the HRI a precision-chop machine is used and the herbage is cut to a length of approximately 20 mm with a resultant high-quality silage. From the work of Dulphy & Demarquilly (1973 & 1975a & b) and Butler *et al*. (1977) it is clear that silage made with a precision-chop forage harvester gave higher animal production than silage made with flail harvester or double-chop machines, but in farm practice the relative capital costs of the harvesting machine and tractors have to be considered. In general, as forage harvesters increase their protential for fineness of chopping their cost and power requirement increase. The ultimate choice of machine depends, therefore, on a compromise between length of chop, capital cost, output per h and silage quality. Wherever possible a machine that chops as short as possible, e.g. approximately 20 mm, is to be preferred, but mechanical and management factors must be considered as well as any anticipated effect on animal production.

## FILLING AND SEALING

The prime aims at this stage of silage making are rapid filling, adequate consolidation and effective sealing. Normally, with clamp silos, the herbage is tipped from the silage trailers and then finally moved by either a buck-rake or similar machine from the pile to the silo. The trailers should, of course, be part of an efficient and integrated silage system to ensure that the transport facilities do not limit performance in the field or at the silo. Trailers of inadequate size may often cause delays and trailer capacity should be examined critically. Small trailers are easy to handle and require relatively low-powered tractors, but more loads per silo have to be transported than with large trailers. Load weight and

DM capacity are influenced greatly by the chop length of the herbage, and the degree of wilting of the crop. For example, the weight of DM in a trailer may be more than doubled if the length of chop of the crop is reduced by altering the number of knives in the precision-chop harvester. Good organization with trailers, and skilled drivers, can do much to improve the rate at which the silo is filled. At all stages of silage making it is important to keep soil out of the crop, but it is particularly vital to avoid transporting mud into the silo at the time of filling. The chopped herbage should be unloaded onto a clean, well-drained area of concrete and the tractor tyres kept as free from mud contamination as possible.

Whatever the system of silage making, the silo should be filled as rapidly as possible. This will help to minimize losses in the field and reduce the length of time the crop is exposed to air in the silo. The latter will help to exclude oxygen from the ensiled material and, as a result, the temperature of the silage will be kept low. Filling should be continuous with as few breaks as possible, and if the crop is chopped short and an additive is used there need not be long delays for extra rolling to aid consolidation. Good consolidation and the exclusion of air are important, and can be achieved by the tractor and buck-rake running over the crop as filling proceeds. When filling a clamp silo it is preferable to first fill the back with a triangular wedge of material and then place sloping layers of herbage on top of this until the silo is filled completely. The wedge system ensures that the minimum area of ensiled material is exposed to the air and that consolidation is achieved in the silo from the earliest stages of filling. The system of putting a thin layer of herbage over the whole area of the silo floor and building up parallel layers of material is not recommended.

Buck-rakes may be either front or rear mounted on the tractor and the types with a mechanism for pushing off the grass are particularly valuable. These relatively simple but important machines have a high work-rate, especially with short-chopped herbage, and are not expensive. For high output systems, large industrial loaders and even crawler tractors may be used. Blowing equipment, which is essential for filling tower silos, is rarely used for clamp silos and in most situations the tractor-mounted buck-rake is the ideal piece of machinery.

During filling, a sheet of plastic should be placed temporarily over the silo each night to prevent heated air rising from the clamp and cold air being drawn in; if air movement is restricted, losses are reduced. The sheet may be held in place overnight by a few discarded motor tyres.

The final sealing of the clamp is a vitally important task, and must be done with great care and precision. The object is to completely exclude air and rain, and thus at the outset the silo walls should be completely air-tight, without cracks. Before filling starts, the walls should be inspected carefully, and any holes and gaps filled with either cement or a plastic sealing compound. The top seal of the silo should consist of one or preferably two layers of 125-$\mu$m (500-gauge) black plastic sheeting. If two sheets are used, the new clean sheet should be placed next to the silage with the old sheet on top. Where sheets are joined, there should be generous overlaps; air can be excluded if the two edges are sealed with a wide adhesive tape. At the junction of the walls and the sheet, i.e. the 'shoulder' of the silo, it is helpful if a separate sheet of plastic is attached to the silo walls before filling and then placed under the top sheets of plastic before taping them together. This simple technique will effectively prevent air escaping from the sides of the silo. To prevent air entering the silo at the front it is worthwhile to bed the plastic sheets to the ground with soft bags of sand weighed down with weights, such as old railway sleepers. The top sheets must be firmly held in close contact with the top of the silage by suitable weights such as tyres or bales of straw. These should cover the entire area of the silo and not be placed merely at odd points. Experiments with a range of surface weights on the plastic sheet indicated that 25-50 kg. $m^{-2}$ reduced the depth of surface spoilage of silage to less than 10 mm, but 100 kg. $m^{-2}$ was needed generally to prevent all visible waste (Hastings, 1976). Car tyres exert a pressure of approximately 25 kg. $m^{-2}$ whereas a layer of straw bales averages approximately 50 kg. $m^{-2}$. Thicker sheets of plastic will also reduce surface waste but are heavy to handle and to move. The sealing practices described above using two sheets of 125-$\mu$m (500-gauge) plastic and old tyres have been used successfully at the HRI for many years, with an absolute minimum of wastage of silage at the surface and sides of the silo. Where silage-making practice is good, with effective sealing throughout the storage period, there should be almost no waste on the silage surface and the losses during storage may be less than 0.10 of the food ensiled. Surface waste indicates poor sealing. With good sealing, the digestibility of the herbage ensiled and the silage should correspond closely, as illustrated by average values of 706 and 712 g. $kg^{-1}$ DM digestibility respectively for 15 comparisons summarized by Flynn (1979). When a clamp silo is well sealed, it is virtually air-tight, resembling closely a vacuum silo that is completely enclosed in plastic. Large vacuum silos are not practical for farm use but the principle they embody may be achieved with a clamp silo by good construction of the floor and walls, and careful sealing of the top.

## TEMPERATURE IN THE SILO

As changes have been made in the silage-making process over the last 20-30 years, the maximum temperature in the silage has tended to fall. To ensure a good fermentation and palatable silage it used to be normal practice to allow the ensiled material to heat up to approximately body temperature, 37°C. At one stage, the filling of the silo was even started on either a Thursday or a Friday to allow the material to warm up before filling recommenced on the following Monday morning. A high temperature is not now recommended, and most of the high-quality silages made in clamp silos at the HRI in the last 6 years have had maximum temperatures below 30°C, with a mean temperature of 23°C. The lowest recorded maximum temperatures were 18 and 19°C in grass silages that had a pH value of 3.8 and DOMD values of 670 and 700 g. $kg^{-1}$, and 16°C in a high-quality red clover silage (Castle & Watson, 1974). Silages made with a cold fermentation may be highly acceptable to cattle, provided there has been a satisfactory silage fermentation. A cold, wet silage with a butyric fermentation, and containing excessive protein breakdown products such as $NH_3$, would not be acceptable but this type of silage may be avoided by application of good ensilage techniques. Low temperatures indicate a low level of metabolic activity within the silage and a limited degradation of nutrients. This was seen in the comparisons between silages either treated with formic acid or left untreated as controls. The maximum temperature of the acid-treated silages averaged 24°C compared with 32°C in the untreated silages, and the low-temperature silages contained 9 g. $kg^{-1}$ more DM and 31 g. $kg^{-1}$ more DOMD (Castle, 1975). Maximum temperatures of 20-25°C in high-quality silages are therefore suggested as target values.

## EFFLUENT PRODUCTION

Silage effluent is a source of DM loss, which may be 60-70 g. $kg^{-1}$ of that in the original herbage. In addition, it is a noxious pollutant, which has extremely harmful effects on rivers and streams. It is thus desirable to make silage that will produce little or no effluent. The amount of effluent produced from silage is influenced primarily by the DM concentration of the crop at the time of ensiling. For example, herbage ensiled with less than 150 g DM per kg will produce approximately 180 l effluent per t whereas that ensiled at 200-250 g DM per kg will produce approximately 20-75 l. $t^{-1}$ (Ministry of Agriculture, Fisheries and Food, 1978). From data obtained at the HRI and from the Grassland Research Institute it was calculated that herbage with a minimum DM concentration of 247 g. kg$^{-1}$ was required to ensure that no effluent was produced from the silo (Castle & Watson, 1973b). However, from the results of Bastiman (1977) it would appear that a little effluent may continue to be produced from herbages with up to 300 g DM per kg. Most of this effluent is produced within the first 3 weeks after ensiling the herbage and the amount is increased by the use of formic acid as an additive. Short chopping and excessive consolidation also increase effluent production, and there are thus additional reasons for aiming to ensile a crop with approximately 250 g DM per kg. Effluent must be collected in a tank constructed of non-porous material but this is an extra capital expense and there is the additional cost of distributing the effluent on the land. DM losses in the effluent increase as the DM concentration of the ensiled crop decreases, and this again emphasizes the need to avoid effluent production from the silo. Thus, for a number of reasons affecting both silage fermentation and effluent production, there are advantages from ensiling herbage with a DM concentration of approximately 250 g. kg$^{-1}$. At the HRI the last 11 silos of high-quality silage have been filled with material containing 261 g DM per kg and effluent production has been negligible.

## EMPTYING SILOS

Although it is widely recognized that losses in silage making may be increased by management factors that allow air into the silo, it has been appreciated only recently that similar losses may occur at the time the silo is emptied. These losses are due to 'secondary fermentation', which is aerobic fermentation occurring as the silo is opened and silage is exposed to the atmosphere. Spoilage and DM losses of 50-100 g. kg$^{-1}$ may occur at this stage. Losses of 59 g DM per kg in grass silages were reported by Crawshaw & Llewelyn (1978) in three unstable silages, but these workers could not predict the degree of stability from the chemical analysis of the silages. Maize silage appears to be particularly vulnerable to losses of DM due to secondary oxidation.

The exposed surface-area of a silo should always be kept to a minimum, with a cleanly-cut face and the minimum disturbance of the silage. If self-feeding is being practised, the settled height of the silage should not exceed approximately 2.3 m for Friesian cattle, and the width of the feeding face should be approximately 200 mm per cow. This width may be reduced slightly, provided the cows have 24-h access to the silo and there is no indication that they are being restricted in their intake. If a uniform silage has been made, and the feeding barrier is adjusted and moved carefully, an even, vertical silage face should result and losses due to oxidation will be minimal. Care in silage making to ensure a uniform silage may reduce losses at the time of feeding.

If silage is being removed from the silo for feeding elsewhere, it is particularly important to maintain an even, tidy silage face. Silage cutters, mounted on a tractor three-point linkage or a foreloader, will remove a clean block of silage weighing approximately 500 kg. Since the silage in the block is practically undisturbed, it is slow to deteriorate and for the same reason the silage remaining in the clamp keeps better because there is little entry of air. If silage is removed from the face with a foreloader in a careless way, an uneven and ragged surface will result, and losses of DM will be high. Experimentally, formic, propionic, acrylic and hexanoic acid have been added to silage, and shown to be effective in reducing losses due to secondary oxidation.

## SUMMARY

From experimental studies and practical experience over the last 10 years it is concluded that good grass silage, with a low loss in the silo, can be made if the following points are carefully attended to.

1. Use a perennial ryegrass and preferably a late variety.
2. Cut early, before ear emergence.
3. Cut when dry, wilt for up to 24 h and do not exceed a DM concentration of approximately 250 g. $kg^{-1}$.
4. Chop to lengths of approximately 20 mm.
5. Apply an effective additive.
6. Fill the silo quickly and evenly.
7. Consolidate the grass tightly.
8. Keep the temperature low, at approximately 20-25°C.
9. Seal the top and edges of the silo with plastic sheet, and cover with tyres or bales.

## REFERENCES

BASTIMAN, B. 1977 Factors affecting silage effluent production. *Experimental Husbandry* No. 31 pp 40-46

BUTLER, T.M., GLEESON, P.A. & COMERFORD, P.J. 1977 Dairy cow nutrition. *Research Report, An Foras Talúntais, Animal Production* 1975, pp 102-104

CASTLE, M.E. 1975 Silage and milk production. *Agricultural Progress* **50** 53-60

CASTLE, M.E. & WATSON, J.N. 1969 The effect of level of protein in silage on the intake and production of dairy cows. *Journal of the British Grassland Society* **24** 187-192

CASTLE, M.E. & WATSON, J.N. 1970a Silage and milk production, a comparison between grass silages made with and without formic acid. *Journal of the British Grassland Society* **25** 65-70

CASTLE, M.E. & WATSON, J.N. 1970b Silage and milk production, a comparison between wilted and unwilted grass silages made with and without formic acid. *Journal of the British Grassland Society* **25** 278-284

CASTLE, M.E. & WATSON, J.N. 1971 Silage and milk production. A comparison between a diploid and a tetraploid ryegrass. *Journal of the British Grassland Society* **26** 265-270

CASTLE, M.E. & WATSON, J.N. 1973a Silage and milk production. A comparison between wilted grass silages made with and without formic acid. *Journal of the British Grassland Society* **28** 73-80

CASTLE, M.E. & WATSON, J.N. 1973b The relationship between the DM content of herbage for silage making and effluent production. *Journal of the British Grassland Society* **28** 135-138

CASTLE, M.E. & WATSON, J.N. 1974 Red clover silage for milk production. *Journal of the British Grassland Society* **29** 101-108

CASTLE, M.E., RETTER, W.C. & WATSON, J.N. 1977 Silage and milk production, a comparison between additives for silage of high digestibility. *Journal of the British Grassland Society* **32** 157-164

CRAWSHAW, R. & LLEWELYN, R.H. 1978 The aerobic deterioration of grass silages. *Proceedings, 5th Silage Conference* (Ed. R.D. Harkess) pp 4-5 Hannah Research Institute, Ayr

CZERKAWSKI, J.W., WATSON, J.N. & BRIGGS, J. 1977 Be safe and sure with silage additives. *Dairy Farmer* **24**(11) 27-31

DULPHY, J.-P. & DEMARQUILLY, C. 1973 [Effect of type of forage harvester and chopping fineness on the feeding value of silages.] *Annales de Zootechnie* **22** 199-217

DULPHY, J.P. & DEMARQUILLY, C. 1975a [Influence of the type of forage harvester on the value of grass silages for heifers of dairy breeds.] *Annales de Zootechnie* **24** 351-362

DULPHY, J.P. & DEMARQUILLY, C. 1975b [Influence of the type of forage harvester on the silage intake level and the performances of dairy cows.] *Annales de Zootechnie* **24** 363-371

FLYNN, A.V. 1979 Conservation of forage for beef cattle. *Grange Cattle Production Seminar,* Paper No. 8. An Foras Talúntais, Grange, Co. Meath

GRANT, A.D. 1978 Quality silage making — a survey 1975 and 1976. *Research and Development Publication* No. 2. West of Scotland Agricultural College, Ayr

HASTINGS, M. 1976 Aspects of harvesting, storage and unloading in practical silage systems. *Proceedings, 4th Silage Conference* Session No. 1, Paper No. 3 Grassland Research Institute, Hurley, Maidenhead

McDONALD, P. 1976 Trends in silage making. In *Microbiology in Agriculture, Fisheries and Foods* (Eds F.A. Skinner & J.G. Carr), *Society for Applied Bacteriology Symposium Series* No. 4 pp 109-121 Academic Press, London

MARSH, R. 1978 A review of the effects of mechanical treatment of forages on fermentation in the silo and on the feeding value of the silages. *New Zealand Journal of Experimental Agriculture* **6** 271-278

MARSH, R. 1979 The effects of wilting on fermentation in the silo and on the nutritive value of silage. *Grass and Forage Science* **34** 1-9

MINISTRY OF AGRICULTURE, FISHERIES AND FOOD 1978 *Silage Making. Profitable Farm Enterprises, Booklet 9* Agricultural Development and Advisory Service

MORRISON, I.M. 1972 A semi-micro method for the determination of lignin and its use in predicting the digestibility of forage crops. *Journal of the Science of Food and Agriculture* **23** 455-463

STEVENSON, M.H. & UNSWORTH, E.F. 1978 The mineral composition of some Northern Ireland silages conserved during the 1976 season. *Record of Agricultural Research, Ministry of Agriculture for Northern Ireland* **26** 11-15

STIRLING, ANNA C. & WHITTENBURY, R. 1963 Sources of the lactic acid bacteria occurring in silage. *Journal of Applied Bacteriology* **26** 86-90

WALDO, D.R., KEYS, J.E., Jr, SMITH, L.W. & GORDON, C.H. 1971 Effect of formic acid on recovery, intake, digestibility, and growth from unwilted silage. *Journal of Dairy Science* **54** 77-84

WALTERS, R.J.K. 1976 The field assessment of digestibility of grass for conservation. *A.D.A.S. Quarterly Review* No. 23 pp 323-328

WILKINS, R.J., WILSON, R.F. & COOK, J.E. 1974 Restriction of fermentation during ensilage. — The nutritive value of silages made with the addition of formaldehyde. *Proceedings, 12th International Grassland Congress,* Moscow, Vol. 3 Part 2 pp 674-690

WILKINSON, J.M., WILSON, R.F. & BARRY, T.N. 1976 Factors affecting the nutritive value of silage. *Outlook on Agriculture* **9** 3-8

WILSON, R.F. & WILKINS, R.J. 1973 Formic acid as a silage additive. I. Effects of formic acid on fermentation in laboratory silos. *Journal of Agricultural Science* **81** 117-124

WILSON, R.F., WOOLFORD, M.K., COOK, J.E. & WILKINSON, J.M. 1979 Acrylic acid and sodium acrylate as additives for silage. *Journal of Agricultural Science* **92** 409-415

## Chapter 6

# Feeding high-quality silage

**M.E. Castle**

*Hannah Research Institute, Ayr KA6 5HL*

Many feeding experiments have been conducted over the years with silages of low and moderate quality but until recently there has been a lack of information about high-quality, highly digestible silages. In the past decade a series of experiments with silages of this type has been undertaken with dairy cows at the Hannah Research Institute (HRI) and the findings of these experiments are summarized here, together with relevant information from other studies.

## Conduct of feeding experiments at the Hannah Research Institute

In all the experiments at the HRI the silage was removed from the face of the clamp with a tractor and foreloader, transported to a shed for weighing into baskets and then offered to the cows. By removing silage from the narrow face of the silo every 2 or 3 d, aerobic deterioration was kept to a minimum and wastage at the silo was negligible.

The silage was offered to the cows three times per d at 07.00, 11.00 and 16.00 h in individual troughs. The amount offered was sufficient to ensure that approximately 100-150 g. kg $^{-1}$ of the original weight was available as a residue for weighing before the next feed. On this system the silage was given virtually *ad lib.* and the cattle had access to the material for approximately 21 h each d, i.e. a time similar to that on many commercial farms. With three weighings of silage and three refusals per day it was possible to keep a close check on intake, and to notice at once any large refusals of food. The technique also ensured clean troughs three times per day and hence allowed the concentrates to be offered in the same trough when it was empty of silage. Both silage and silage refusals were weighed with an accuracy of 0.5 kg.

The construction of the cow-stall and the trough (plate 1) was such that adjacent cows could not steal each other's silage and, although a little silage dropped between the trough and the cow's fore-feet, this wastage was readily collected and weighed with the refusals. Occasionally, individual cows tossed silage into the air, but waste was prevented

**Plate 1. Individual cow stall used in feeding experiments.** Note high partitions, water bowl outside trough and rubber mat with sawdust

by putting wire netting over the top of the stall divisions. When it was not possible to give minerals in the concentrates, the mineral mixture was scattered on the silage prior to feeding. This technique was successful although the intake of the mineral may not have been as complete as when the mixture was included in the concentrates. Throughout all the studies, the silage on offer was weighed daily for each individual cow, whereas refusals were taken on 5 consecutive d per week. The individual feeding of cows with silage in this way has given accurate results, with low standard errors, using small numbers of cows (Castle *et al.*, 1977c).

## SILAGE FOR MILK PRODUCTION

### Diets of silage alone

Using the facilities and techniques described above, a wide range of voluntary intakes of silages has been observed in the feeding experiments undertaken. In some experiments, silage was the sole constituent in the ration, and the various intakes recorded are listed in table 1. In these studies the Ayrshire cows were normally approximately 8 weeks post-calving at the start of the experimental feeding period. The daily intake of silage dry matter (DM) ranged from 10.4 to 12.8 kg per cow, with a mean value of 11.3 kg. This was equivalent to 24.1 g. $kg^{-1}$ mean live weight but, for the period of mid-lactation, is clearly below the maximum intake that could be achieved with a mixed diet. Silage is rarely used as the sole food for dairy cows and no comparable data on intake have been found in the literature for rations of silage only. It has been shown that mature fattening cattle may achieve high rates of live-weight gain on good-quality silage alone but gains with younger cattle on silage alone have been disappointing (Forbes & Jackson, 1971). In the results given in table 1 the mean daily milk yield was 14.4 kg per cow, which indicates the milk-producing potential of this type of high-digestibility silage. However, as discussed by Castle *et al.* (1977c), the milk yields from the 1973, 1974 and 1975 silages were associated with a daily live-weight loss of 0.70 kg per cow and, if this is allowed for, the silage was providing nutrients for maintenance and the production of 10.8 kg milk per d. The daily losses in live weight with the 1977 silages averaged 0.48 kg per cow but in all the experiments some of the losses in live weight were probably due to changes in gut-fill. The silage-only rations supplied, on average, 1.12 of the metabolizable energy (ME) and 1.06 of the digestible crude protein (DCP) required theoretically for maintenance and milk production, and it is not easy to identify the limiting nutrient. DM intake appears to provide the ultimate limitation on production and this highlights the importance of the factors discussed earlier that influence silage intake. If, as with mixed hay and concentrate diets, intakes of 29-30 g. $kg^{-1}$ live weight could be achieved, daily yields of 20 kg milk per cow could theoretically be obtained from diets of silage alone. For example, with the 1977 silage, containing 712 g digestible organic matter in the dry matter (DOMD) per kg, three cows on the experiment had an average silage intake equivalent to 30.2 g. $kg^{-1}$ live weight and yielded 19.7 kg milk per day with a small gain in live weight. This isolated observation with a small number of animals may not be typical but it serves to illustrate the potential of silage in the diet of the lactating cow. This particular silage (table 1) produced milk with the highest crude protein content of all the silage-only rations used in recent years.

## Supplementary foods for use with silage

### *Effects on silage intake*

As indicated above, a diet of silage alone will limit the intake of DM and hence milk yield, and thus silage is usually supplemented with other foods. These are normally concentrates, often high in energy, and in theory a food such as barley should be adequate in most situations. For example, if silage intake was 25 g. $kg^{-1}$ live weight, supplementation with a barley-based concentrate to the theoretical limit of appetite (Ministry of Agriculture, Fisheries and Food *et al.*, 1975) should give an energy intake sufficient for maintenance and the daily production of almost 30 kg milk per cow. In practice, however, this does not occur, as most supplements reduce silage intake. The size of this reduction depends on the amount and, in particular, the type of supplement. To calculate the exact effect of the supplement it is necessary to have a control diet of silage only in the experiment, although the general trend may be seen in some trials without a control.

Barley has a particularly marked effect on silage intake, and in four comparisons the mean reduction in silage DM intake was 0.51 kg. $kg^{-1}$ of barley DM (Castle & Watson, 1975 & 1976) over a range of 3.3-6.0 kg. $d^{-1}$ barley DM per cow. A reduction of 0.64 kg silage DM per kg barley DM was found by Ettala & Lampila (1978) in 13 trials with 296 Ayrshire cows, with a mean barley DM intake of 2.1 kg. $d^{-1}$. This reduction is attributed to a reduction in the ruminal digestibility of the silage because of the effect of the cereal starch on the rumen microflora (see Chapter 4). It had been postulated that an increase in the number of individual meals of barley per 24 h might alter favourably the reduction

in silage intake but this has not occurred. In a recent experiment (M.S. Gill & M.E. Castle, unpublished results) with lactating cows, silage intake was not altered significantly when a barley cube was offered, 2, 4 and 22 times per 24 h.

Dried grass cubes have an effect similar to barley on silage intake but the reduction is smaller, averaging 0.36 kg silage DM per kg grass DM, and tends to be highest at the lowest rate of dried-grass intake (Castle & Watson, 1975).

Molassed beet pulp has been found to have a slightly smaller effect on silage intake than barley and in a recent experiment, in which some soya bean meal was offered on all treatments, the reduction in silage DM intake was 0.40 kg. $kg^{-1}$ beet pulp DM compared with 0.44 kg. $kg^{-1}$ barley DM.

In contrast, the feeding of oilseed meals does not reduce silage intake. In three separate experiments with groundnut cake the mean intake of silage DM was increased by 0.13 kg. $kg^{-1}$ supplement DM offered, up to a daily maximum of 3.4 kg groundnut per cow. A typical result is shown in table 2. Groundnut cake also partially offsets the adverse effects of the barley (table 2); a reduction in silage DM intake of 0.32 kg. $kg^{-1}$ DM was observed with a supplement mixture of 830 g barley and 170 g groundnut per kg (Castle & Watson, 1976). This suggests that the protein content of the supplement has a beneficial effect on silage intake in addition to the effect that would result from the supplement having a reduced starch content. Soya bean meal has an effect similar to groundnut cake and has increased silage DM intake by 0.06 kg. $kg^{-1}$ soya DM (Retter, 1978). In another experiment, the daily silage intake was increased from 7.5 to 8.4 kg DM per cow when groundnut was replaced by an equal weight of soya (Castle & Watson, 1979) and this suggests that soya is preferable to groundnut as a supplement for silage. The degradability of the protein in soya is less than that in groundnut and this is a possible explanation for the different effect of the two protein foods.

Hay is sometimes used as a supplement for silage and this practice has been investigated in a recent series of experiments at the HRI (Retter, 1979). In eight separate comparisons, the mean reduction in silage intake was 0.84 kg DM per kg hay DM. This reduction occurred with hays with DOMD values ranging from 580 to 700 g. $kg^{-1}$ and with silages varying from 598 to 683 g. $kg^{-1}$. The level of reduction was unaffected by the addition of soya bean meal or groundnut meal to the ration and it was concluded that long hay, regardless of its DOMD content, was of little value as a silage supplement. In a further experiment comparing high-quality hay in either the long or the ground and cubed form, the total intake of silage plus hay was increased by grinding and cubing the hay. However, this process involves additional cost, and the technique is unlikely to be worthwhile on a commercial scale when high-quality silage can be offered simply and cheaply to the dairy herd. In studies in Finland, Ettala & Lampila (1978) have also found hay to have a high replacement value when used as a supplement for silage; silage DM intake was found to be reduced on average by 1.15 kg. $kg^{-1}$ hay DM.

A summary of the effect of supplements on silage intake is given in table 3.

**Table 3. Changes in the intake of silage dry matter with different supplementary foods**

| Supplementary food | Change in silage intake (kg DM per kg supplement DM)† |
|---|---|
| Hay | − 0.84 |
| Barley | − 0.51 |
| Dried grass cubes | − 0.36 |
| Barley + groundnut | − 0.32 |
| Sugar beet pulp | − 0.40 |
| Soya | + 0.06 |
| Groundnut | + 0.13 |

† − Denotes a decrease and + an increase in intake

## *Effects on milk production*

As the supplements affected the amount of silage eaten and hence the total intake of nutrients, there were large differences between supplements in their effects on milk yield. The poorest responses in milk yield occurred with supplements solely of barley, with daily yields ranging from 15.7 to 17.0 kg per cow compared with a mean yield of 14.4 kg on the control rations of silage alone (Castle & Watson, 1975 & 1976). A typical effect is seen in table 2. The poor response in milk production to supplementation of the diet was due partly to the reduced intake of silage but there was also an apparent inefficiency in the use of the energy for milk production. For example, in one experiment (Castle & Watson, 1976) the ME intake on the silage and barley ration was 1.46 of the theoretical requirements for maintenance and milk production, and yet the milk-yield response was lower than on other treatments where the ME intake was only 1.09 of the theoretical requirement. Expressed in another way, the apparent efficiency of utilization of ME for lactation was only 0.39 for barley plus silage compared with 0.46 for silage alone, and

0.54 for groundnut plus silage (Thomas & Castle, 1979). This suggests that milk-yield responses due to increased energy intake and those due to increased apparent efficiency of utilization of ME may occur independently of each other. Thus, substitution of groundnut for a portion of the barley in the diet (table 2) increased silage intake, total energy intake, the efficiency of energy use and hence milk yields. A similar effect was seen in a larger-scale experiment when concentrates containing 140 and 180 g crude protein per kg were compared as supplements to a silage diet (Laird *et al.*, 1979). On the treatment containing the high-protein concentrate, the intake of silage was increased by 0.10 and the daily milk yield rose from 17.4 to 19.5 kg per cow. The same trend was seen in the study of Gordon (1979a) when an increasing protein content in the concentrate resulted in increasing yields of milk. The cows had access to a grass silage with a DOMD value of 670 g. $kg^{-1}$, containing 140 g crude protein per kg DM, and the increases in milk yields due to protein treatment were linear (table 4). Further studies by Gordon & McMurray (1979) suggested that maximum milk output per cow would be achieved with a concentrate supplement containing 244 g crude protein per kg fresh weight. There is also evidence that the response in milk yield to protein supplements is greater with high-digestibility silage that has been wilted, rather than not wilted (Gordon, 1980).

The general effect of substituting a protein concentrate for barley

**Table 4. Silage intake and milk yield with supplementary concentrates of different protein concentrations (Gordon, 1979a)**

| | Protein concentration of concentrate (g. $kg^{-1}$ fresh weight) | | | | s.e. of mean |
|---|---|---|---|---|---|
| | 95 | 137 | 174 | 209 | |
| **Silage intake (kg DM per cow per d)** | 6.80 | 7.50 | 7.80 | 7.40 | 0.32 |
| **Milk yield (kg per cow per d)** | | | | | |
| Mean | 18.0 | 19.3 | 20.4 | 21.7 | 0.54 |
| Peak | 19.7 | 21.2 | 22.2 | 23.9 | 0.59 |
| **Milk composition (g. $kg^{-1}$)** | | | | | |
| Fat | 39.9 | 38.6 | 37.7 | 37.6 | 0.90 |
| Protein | 30.2 | 31.1 | 31.7 | 31.2 | 0.41 |
| Lactose | 46.1 | 45.5 | 46.5 | 45.6 | 0.34 |
| Solids-not-fat | 86.9 | 88.0 | 88.5 | 87.7 | 0.42 |

when feeding a grass silage (Castle & Watson, 1969) has occurred also with red-clover silage with a DOMD value of 572 g. $kg^{-1}$ (Castle & Watson, 1974). With this type of silage, total DM intake and milk yield were increased when groundnut cake was substituted for an equal weight of barley.

As discussed previously, dried grass had less effect than barley in reducing the voluntary intake of silage (Castle & Watson, 1975) and milk yield with dried-grass supplements was consistently higher than with barley supplements. The apparent efficiency of utilization of ME for lactation averaged 0.62 for dried grass and silage diets compared with 0.46 for three barley and silage diets (Thomas & Castle, 1979). Silage intake was increased significantly when dried grass was included in conventional cereal-based concentrates (Tayler & Aston, 1973; McIlmoyle *et al.*, 1975) and it has been suggested that this increase compensates for the fact that the dried grass is generally lower in nutritive value than the concentrates. Also, Gordon (1975) showed that increasing the proportion of dried grass in the concentrate with a diet including silage *ad lib.* did not reduce milk yield or alter milk composition. In general, it may be concluded that ground and cubed dried grass is an excellent supplement for dairy cows given silage; daily milk yields of over 20 kg per cow were recorded by Castle & Watson (1975) from animals offered high-quality silage *ad lib.* plus dried grass at the rate of 0.3 kg. $kg^{-1}$ milk.

Responses in milk yield to supplements of hay in cows given silage diets have been examined in a number of experiments (Retter, 1978). As discussed previously, hay is a poor supplement with a large depressive effect on silage intake and, consistent with this, hay given long, chopped or ground and cubed had no significant effect generally on milk yield in cows given high-quality silage. In one experiment, where the intake of silage was low, a supplement of hay with a DOMD value of 650 g. $kg^{-1}$ gave a significant increase in milk yield but with silage of higher quality the effect was not apparent. Thus, if silage is made under satisfactory conditions, the use of hay will not be justified but, if conditions are adverse and silage has poor intake characteristics, there may be a limited place for supplements of hay of high quality. In commercial practice, the making and feeding of hay is an extra operation, and the machinery and labour costs involved should be considered also.

In marked contrast to the hay, the effect of the protein supplement is invariably to increase milk yield. In one experiment, for example, the average daily yield of milk was increased significantly, from 14.0 kg per cow on a silage-only ration to 18.1 kg per cow when a supplement of

2.3 kg of groundnut DM was offered with the silage. In this experiment, the supplement of groundnut increased the voluntary intake of silage and thus there was a net increase in both ME and DCP intake on the groundnut treatments (Castle *et al.*, 1977b). Extra energy and protein were supplied by the supplement but, equally important, there was no reduction of nutrient intake via the silage part of the ration.

## *Silage 'balancer' cubes*

Experiments of the type just described raise the interesting possibility that a useful supplement for high-quality silage would be a concentrate mixture of high protein content and with a low barley content. With this combination of characteristics there should be the maximum intake of silage coupled with the minimum weight of the supplement. A supplement of groundnut only (481 g DCP per kg DM), given at the rate of 0.1 kg. $kg^{-1}$ milk with high-quality silage, did not depress silage intake, and the total ration supplied 1.09 and 1.43 of the theoretical requirements of ME and DCP respectively (Castle & Watson, 1976). With a cube containing 82.2 g groundnut cake, 5.0 g molasses and 12.8 g minerals per kg, the optimum rate of feeding appeared to be 0.14 kg. $kg^{-1}$ milk (Castle *et al.*, 1977c). This cube was designed to supply an adequate amount of Ca, $P_2O_5$ and MgO, and the incorporation of minerals reduced the DCP concentration to 329 g. $kg^{-1}$ DM. Another high-protein cube (265 g DCP per kg DM), used in an experiment on a commercial farm, contained 650 g groundnut, 180 g barley, 50 g molasses and 120 g minerals per kg (Castle *et al.*, 1977a). This cube was given at the rate of 0.15 kg. $kg^{-1}$ of milk with high-quality silage and 9 kg of brewers' grains per cow daily, and gave milk yields similar to a barley mix offered at 0.4 kg. $kg^{-1}$ milk.

It should be stressed that in these three experiments the silages had high DOMD values and they were thus capable of supplying adequate energy to the cows. The 'balancer' cube, because it was given at a low rate, was not a major source of energy, and the cube should be regarded as supplying primarily protein and minerals. This aspect of the cube was demonstrated in a comparison between a high-protein cube (360 g crude protein per kg) given at 0.15 kg. $kg^{-1}$ milk and a normal-protein cube (180 g crude protein per kg) given at 0.3 kg. $kg^{-1}$ milk, when the daily milk yields were 15.6 and 17.1 kg per cow respectively (Laird *et al.*, 1981). The different feeding treatments started shortly after calving, and the silage with a mean DOMD value of 622 g. $kg^{-1}$ was the limiting factor in supplying energy, and hence in limiting milk production, on the high-protein concentrate treatment offered at a low rate. Clearly, if the 'balancer' cube is not going to be an important source of energy, it must be formulated to encourage the maximum intake of high-quality silage. Also, if the silage is not of the highest quality, the balancer cube should be offered only after the peak of lactation, when DM intake is at a maximum, and silage can replace concentrates without any major detrimental effect on milk production.

In all these studies with balancer cubes, the lowered intake of concentrates was compensated for by an increased intake of silage. This increase ranged from 0.11 to 0.19 of the amount consumed on the high-concentrate ration and was equivalent to an extra 1.5-2.0 t silage per cow during the winter feeding period. Thus the use of a high-protein balancer cube could ultimately change the overall management and cropping of the farm so that more silage of higher quality was made, but less barley was grown and fewer concentrates purchased. More development work with balancer cubes is required but the concept has the major attraction of increasing the use of home-produced silage.

## FACTORS AFFECTING THE FEEDING VALUE OF SILAGE

### Digestibility

The experiments conducted at the HRI have indicated clearly that, other things being equal, the DOMD of silage offered *ad lib.* is the major factor influencing milk production. In the experiments summarized in fig. 1 an equal weight of barley concentrates was given with each of a variety of silages. On average, an increase of 10 g. $kg^{-1}$ in the DOMD concentration of the silage increased daily milk yield by 0.24 kg. In all instances silages were offered *ad lib.* and the differences in milk yield were due primarily to differences in voluntary intake between silages. An increase of 0.024 kg milk per g DOMD per kg may seem small, but the cumulative effect is large and stresses the importance of silage quality where maximum use is to be made of conserved forage for milk production. As discussed previously, date of cutting is a prime factor influencing silage digestibility, but minimum wilting, good sealing of the silo and the correct use of additives etc. are needed to maintain silage quality, and hence milk yield. In any silage system it is vital to have sufficient silage but, when that objective can be satisfactorily achieved, it is worth-while aiming for silage with as high a DOMD as possible.

The effect of silage digestibility on milk yield and thus the concentrate-sparing effect of silages with high DOMD may be seen also in two other experiments: In a recent experiment at the HRI, a medium-cut (DOMD =

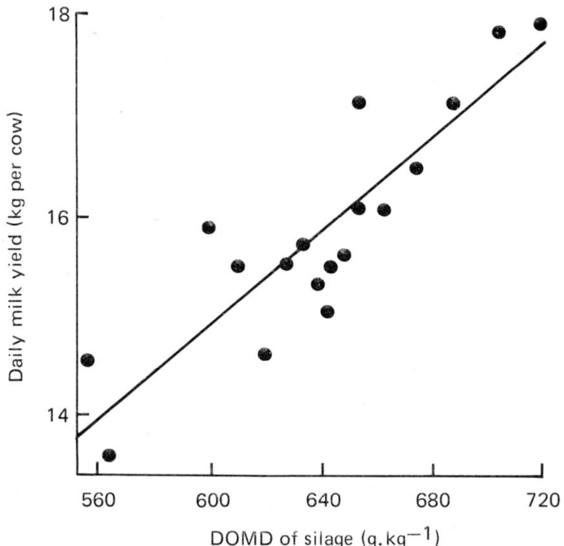


**Fig.1.** **Relationship between the DOMD of grass silage, determined *in vitro*, and the average daily milk yield of the cows eating each silage plus concentrate (Castle, 1975)** (● represents a specific silage)

650 g. kg$^{-1}$) and a late-cut silage (DOMD = 624 g. kg$^{-1}$) required daily concentrate supplements of 2.69 and 2.91 kg DM per cow respectively to give the same daily milk yield as an early-cut silage (DOMD = 712 g. kg$^{-1}$) (Castle *et al.*, 1980). Similarly, in an experiment comparing two silages, Gordon & Murdoch (1978) calculated that 1.9 kg of additional concentrates per cow per day would be required with a medium-quality silage (DOMD = 650 g. kg$^{-1}$) to give a milk output equivalent to that achieved with a higher-quality silage (DOMD = 671 g. kg$^{-1}$). Both these studies show that a high-digestibility silage can make a considerable saving in concentrate use, but the results again underlined the fact that the intake of the high-digestibility silage was always higher than that of the lower-digestibility silage.

## Silage pH

There is conflicting evidence on the possible effects of the pH of silage on intake and milk yield. Occasionally, on commercial farms, silages with low pH values, e.g. 3.5-3.6, have been associated with low intakes and reduced yields of milk, but there is little information experimentally of a link between pH and milk yield. From results with 17 silages in feeding experiments at the HRI there is no clear-cut relationship between pH and milk output but many other factors such as DOMD and DM concentrations confound the issue. The range of pH values in the study was 3.7-5.1 and thus experience with extremely low values is lacking. In studies with sheep it would appear that the intake of well preserved silages may be limited by their low pH (Wilkins *et al.*, 1971), although in a recent survey of 142 silages no significant relationship was found (Wilkins *et al.*, 1978). Incidentally, it was concluded in this study that analysis for ammonia-nitrogen and total N would provide a good guide to the likely relative intakes of different silages. With sheep and young cattle, increases in silage intake resulted from the partial neutralization of silage with sodium bicarbonate (McLeod *et al.*, 1970) but, with dairy cows receiving a silage of pH 3.82, the bicarbonate reduced silage intake (Farhan & Thomas, 1978). In another experiment at the HRI, a supplement of 150 g sodium bicarbonate per d increased silage intake by approximately 9% whereas 300 g. $d^{-1}$ had little effect on intake (M.E. Castle & J.N. Watson, unpublished results). The bicarbonate treatments had no significant effect on milk yield, and it is concluded that if silages with a pH of 3.8 and higher are offered to dairy cows there is little or no evidence to suggest that partial neutralization is worth-while. If silage has a pH below 3.8, then an addition of approximately 25 g sodium bicarbonate per kg concentrate may be worth-while. A low pH is an important part of the silage-making process and its attainment should not be avoided because of fears of its effect on intake.

## Silage chop-length

The general effect of chopping and lacerating herbage prior to ensiling was shown earlier (Chapter 5) to improve silage fermentation. The effect on digestibility seems to be small and variable, although further studies on this topic with cattle are required before firm conclusions may be drawn. From the information reviewed by Marsh (1978), it is clear that, in terms of DM intake, sheep respond well to an increasing fineness of chop in silage but cattle appear much less responsive, due to the frequent practice of feeding concentrates in the cattle experiments. For example, with steers given no supplementary concentrates, Murdoch (1965) found that the DM intake was 22% greater with chopped than with lacerated silage. In contrast, with a dairy concentrate included in the ration, Dulphy & Demarquilly (1975) reported an increase of only

12% in DM intake when silage of 40-mm chop-length was compared with one of 100 mm. In an experiment at the HRI (table 5), in which silage was the sole food of the cows, intake was increased by 33% as a result of chopping the silage short (9 mm) whereas, when the same silage was supplemented with concentrates, the increase due to short chopping averaged 14% (Castle *et al.*, 1979). Milk yield was increased by approximately 7% as a result of reducing chop-length from 72 to 9 mm when adequate protein was in the ration. More recently, no difference in milk yield was found when silages of 12 and 18 mm were compared in rations containing a high proportion of concentrates.

It is thus apparent that there are interactions between silage chop-length, concentrate intake and type of concentrate, and they have a cumulative effect on milk production. Recent studies with cattle in Ireland showed that a decrease in the length of chop from 17 to 6 mm increased DM intake from an average of 7.2 to 8.0 kg per animal per d, whereas the intake with single-chop material was 7.4 kg (Flynn, 1979). Carcass gains on the three types of silage were similar. At present, it is concluded that although there are some nutritional advantages in favour of short-chop silages these may be modified by other factors.

**Table 5. Intake of silage dry matter (kg per cow per day) by cows given silages of different chop-length *ad lib*.**

| Silage length (mm) | Dry cows given silage alone | | Lactating cows given silage and concentrates | |
|---|---|---|---|---|
| Long (72) | 6.97 | (1.00)† | 8.12 | (1.00) |
| Medium (17) | 8.34 | (1.20) | 8.53 | (1.05) |
| Short (9) | 9.24 | (1.33) | 9.28 | (1.14) |

† Values in parentheses give the intake relative to that observed with the long-chopped silage

## Wilting

Almost all the experiments conducted at the HRI with high-quality silages have been with wilted materials and, as has been discussed in Chapter 5, wilting herbage in the field before ensilage has some mechanical and managerial advantages over direct cutting.

The effect of wilting on silage intake has recently been fully reviewed by Marsh (1979), who studied 61 comparisons from 25 sources, and a summary of his findings is presented in table 6. From these results there is no doubt that wilting increased voluntary intake, but it is noteworthy

**Table 6. The effect of wilting on silage intake (summarized from Marsh, 1979)**

| Type of animal | Concentrate feeding | Proportional increase in silage DM intake due to wilting |
|---|---|---|
| Sheep | No | 0.44 |
| Young cattle | No | 0.31 |
| | Yes | 0.12 |
| Milking cows | No | 0.25 |
| | Yes | 0.10 |

**Table 7. The effect of wilting on silage intake, milk yield and efficiency (based on Marsh, 1979)**

| | Silage treatment | | |
|---|---|---|---|
| | Unwilted | Wilted | Proportional change |
| Silage DM intake (kg per cow per d) | 9.1 | 10.0 | + 0.099 |
| Fat-corrected milk yield (kg per cow per d) | 14.2 | 14.6 | + 0.028 |
| Efficiency (kg milk per kg silage DM) | 1.56 | 1.46 | − 0.064 |

that the smallest increase, 10%, was with milking cows receiving concentrates. The response to the wilting in terms of increased DM intake varied from 1-25% in the different experiments with dairy cows but was always positive. In general, the proportional increases in DM intake due to wilting were greater than the proportional decreases in digestibility, and thus it would be reasonable to assume that animal production would be increased as a result of wilting. On average in these experiments there was a slight increase in milk yield of 2.8% due to wilting (table 7) but the effect was not a consistent one and, for example, in the experiment of Castle & Watson (1970) the effect of wilting on milk yield was not significant. In some of the other trials reviewed by Marsh (1979), wilting reduced milk yield when expressed on a fat-corrected basis. Thus, when the efficiency of the silage for milk production is calculated (table 7) there is no advantage with wilted silage when compared with well preserved unwilted silage. More recent experiments reported by Gordon (1980) fully support the above conclusions and include evidence that heavily-wilted herbage may even reduce milk yield. Wilting experiments have also been reviewed and summarized by Flynn (1979), who showed that increased DM intake as a result of wilting was reflected in increased production of milk and meat only when wilting improved silage preservation, and the unwilted material was poorly preserved. In a recent experiment at the HRI (M.E. Castle & J.N. Watson, unpublished results), two unwilted and two wilted silages from the same sward, all receiving additives, were compared. On average, the unwilted and wilted silages had DM concentrations of 200 and 243 g. $kg^{-1}$, had mean intakes of 10.4 and 9.6 kg DM per d, and gave daily milk yields of 19.8 and 17.9 kg per cow respectively. The overall efficiency of utilization of the silages for milk production was slightly lower for the wilted than for the unwilted silages and this difference was shown yet again in a further recently completed experiment at the HRI.

The conclusion, therefore, is that if unwilted silage is well made and has a good fermentation, i.e. made with an effective additive, it is a more efficient forage for milk production than wilted material. Wilting is a simple, but not always fool-proof, means of improving silage fermentation and intake, but it is associated with a slightly reduced efficiency of utilization. As has been said, wilting to approximately 250 g DM per kg has managerial advantages, and may be seen as helping the ensilage process and reducing effluent loss, but it is extremely doubtful whether there will be consistent benefits in terms of animal production.

## FEEDING SYSTEMS

### Allocation of silage and concentrates

From the above discussion it will be apparent that silage should not be regarded as supplying a fixed amount of DM and nutrients to the cow each day throughout lactation. Apart from the physical and chemical characteristics of the silage, such as the concentration of ammonia and DOMD, which affect the basic silage intake, the amount and type of supplementary food has a large effect on silage intake. In addition, if silage is given *ad lib.*, as in most modern feeding systems, the voluntary intake will vary greatly at different stages of the lactation. Furthermore, the silage may be supplemented with concentrates given at either different amounts according to the stage of lactation or at a fixed amount throughout lactation. The former system of concentrate allocation involves the individual rationing of all the cows, whereas the latter system merely demands some simple flat-rate method of feeding groups of animals. Whatever the system of concentrate allocation during the lactation, it is worth-while to offer the cows the silage with the highest DOMD in the first 15 weeks of lactation, or longer if possible. At this early stage of lactation the intake of DM is increasing, from a minimum shortly after calving to a peak, and it is thus important that the food has the maximum nutrient density. Silage with a DOMD of 680-700 g. $kg^{-1}$ is ideal at this stage and, since the intake of ME increases with the concentration of ME in the total diet DM (Leaver, 1980), concentrate feeding is also vital at this period.

If a system of variable concentrate allocation is adopted the concentrate allowance in early lactation may approximate to the conventional 0.4 kg. $kg^{-1}$ milk and, with high-yielding cows, the concentrates may be offered more often than twice daily. In mid-lactation, e.g. from weeks 15 to 30, milk yields will be declining, intake will be at a maximum and thus increasing reliance may be placed on silage for milk production. If high-quality silage is offered, concentrate feeding may be reduced to a level that allows a decline in milk yield of approximately 2.0 - 2.5% per week. Depending on silage quality and intake, the rate of concentrate feeding could be 0.2-0.3 kg. $kg^{-1}$ milk, allowing considerable economies in concentrates as compared with conventional feeding rates. In the last phase of lactation, i.e. after week 30, maximum reliance may be placed on the silage and concentrate feeding reduced drastically or even eliminated. As shown earlier, diets consisting solely of silage with a DOMD of 680 g. $kg^{-1}$ will sustain milk yields of approximately 14 kg. $d^{-1}$. Over the whole lactation, therefore, with a silage of high quality, the amount of concentrates used should average approximately 0.2 kg. $kg^{-1}$ milk.

Variable concentrate allocation to individual animals involves considerable time and effort, but is efficient in optimizing the output from the individual cow and is often favoured for small herds or where there is a wide spread in calving pattern. However, there is now increasing evidence from experimental studies and farm experience that the alternative system of offering a fixed flat-rate of concentrate feeding to all cows may be as effective as individual rationing (Taylor, 1979). The flat-rate system has many attractions with a large herd where groups of cows calve within a relatively short period and may be treated in a uniform way. The work of Østergaard (1979) indicates clearly that a flat-rate pattern of concentrate allocation gives similar yields of milk to more complex systems with a variable pattern of allocation. There are also considerable data from Northern Ireland that show that spring-calved cows (January-February) given silage *ad lib.* give the same milk yield on either a flat-rate or a variable system of concentrate feeding (Steen & Gordon, 1980a & b). The results indicated clearly that milk yield was not affected either when the cows were indoors on the silage or later when they were at grass, and the mean lactation yields of these animals were over 5000 kg. Two points are made by Gordon (1979b). First, no benefit to lactation yield was obtained from trying to improve peak yields by feeding large amounts of concentrates in early lactation; and, secondly, all cows within a herd could be given the same weight of concentrates irrespective of milk-yield potential. These conclusions, based on a series of long-term studies, lead to a simple approach to cow feeding, in which a similar amount of concentrates per day is given to each cow irrespective of lactation number, date of calving or milk yield. In these experiments the silage was offered *ad lib.* and had an ME of approximately 10.0 MJ. $kg^{-1}$ DM. The suggested daily intake of fresh concentrates was 7.2 kg per cow, and the concentrate had an ME of 12.7-12.9 MJ. $kg^{-1}$ DM and contained approximately 180 g crude protein per kg.

A successful system with autumn-calved cows has also been used at the Grassland Research Institute. The silage is given to appetite and concentrates offered at a flat rate irrespective of either the yield potential of the cow or the stage of lactation (Thomas *et al.*, 1978). The silage is given to the cows in troughs plus 7.1 kg concentrate DM supplied daily to each animal. The adoption of this simple technique has been associated with increases in milk yields over the last few years, and a reduction in concentrate use both per cow and per kg milk. In 1976-77 the yield per cow averaged 5678 kg with the use of 1.21 t concentrates per cow, equivalent to 0.21 kg. $kg^{-1}$ milk. This performance demonstrates that a high-quality silage may produce excellent milk yields with only the minimum reliance on concentrated foods. The stocking rate of the herd, at 0.50-0.55 ha per cow, was similar to that achieved on many other intensive farms, but milk yield was higher and concentrate use much lower. A similar simple approach to silage and concentrate feeding is being used by R. Newcombe (personal communication) in Herefordshire. The silage is well made with a DOMD of 650-670 g. $kg^{-1}$ and is self-fed behind an electric wire. All the cows calving in September and October receive 5.5 kg concentrates per day irrespective of milk yield, and this quantity is reduced by 0.9 kg on the 1st d of each month until spring grazing starts. Brewers' grains are given also, but the vital point in the system is the reliance on silage of high quality with only a minimum use of concentrates, i.e. approximately 0.9 t per cow for the winter feeding period.

## Provision of water

The high requirement of dairy cows for drinking water is well recognized but there is sometimes uncertainty about the amounts of drinking water that need be provided for cows consuming high-moisture foods such as silage. The water intake of dairy cows receiving a wide range of silage rations has been recorded in most of the recent feeding experiments at the HRI (e.g. Castle *et al.*, 1977c) and, in addition, the water intake of dairy cows was measured in 14 herds with a total of 840 animals (Castle & Thomas, 1975). Silage was offered to the cows in nine of these herds and it was clear from these studies that the major factors affecting voluntary water intake were milk yield, total DM intake and the DM content of the ration. Water in the food and water drunk are, of course, complementary, and this is shown by the relatively constant ratio of water intake per kg DM consumed. In the farm study this value was 3.70 kg water per kg DM after deducting the weight of water in the milk and this ratio was not affected significantly by any of the variables measured in the study. This agrees well with the estimate of 3.60 kg. $kg^{-1}$ DM given by the Agricultural Research Council (1965) for cattle at an environmen-

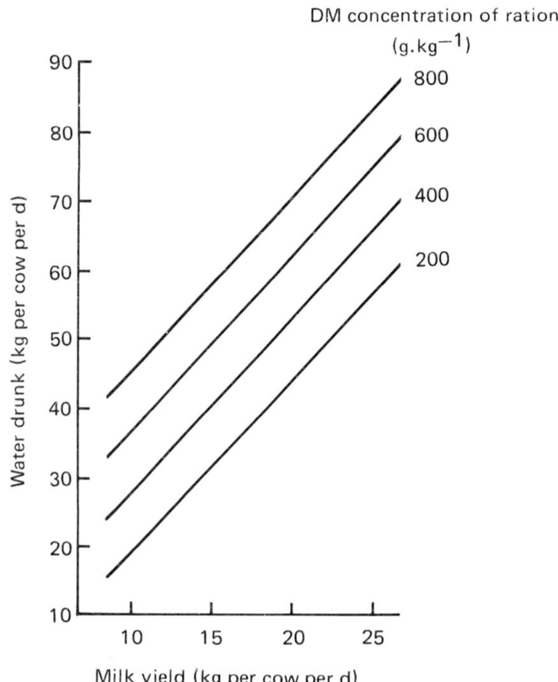


**Fig.2. Relationship between the daily intake of drinking water and the daily milk yield per cow on rations with four different dry-matter concentrations**

tal temperature of 10-15°C. In experiments at the HRI with rations of silage only, the water intake was 3.97 kg. $kg^{-1}$ DM, which was similar to the mean value of 4.04 kg. $kg^{-1}$ DM on rations of silage plus groundnut. When the silages were supplemented with cubed dried grass and barley, total water intake was increased but the ratios of water intakes to DM intakes were reduced to 3.48 and 3.01 kg. $kg^{-1}$ DM respectively.

The requirements of water calculated from the farm data are summarized in fig. 2, and cover a range of milk yields and types of forage, including silage. As a safety measure, 8 kg of water per d should be added to any estimate derived from fig. 2.

## REFERENCES

AGRICULTURAL RESEARCH COUNCIL 1965 *The Nutrient Requirements of Farm Livestock. No. 2, Ruminants* Agricultural Research Council, London

CASTLE, M.E. 1975 Silage and milk production. *Agricultural Progress* **50** 53-60

CASTLE, M.E. & THOMAS, THESCA P. 1975 The water intake of British Friesian cows on rations containing various forages. *Animal Production* **20** 181-189

CASTLE, M.E. & WATSON, J.N. 1969 The effect of level of protein in silage on the intake and production of dairy cows. *Journal of the British Grassland Society* **24** 187-192

CASTLE. M.E. & WATSON, J.N. 1970 Silage and milk production, a comparison between wilted and unwilted grass silages made with and without formic acid. *Journal of the British Grassland Society* **25** 278-284

CASTLE, M.E. & WATSON, J.N. 1974 Red clover silage for milk production. *Journal of the British Grassland Society* **29** 101-108

CASTLE, M.E. & WATSON, J.N. 1975 Silage and milk production. A comparison between barley and dried grass as supplements to silage of high digestibility. *Journal of the British Grassland Society* **30** 217-222

CASTLE. M.E. & WATSON, J.N. 1976 Silage and milk production. A comparison between barley and groundnut cake as supplements to silage of high digestibility. *Journal of the British Grassland Society* **31** 191-195

CASTLE, M.E. & WATSON, J.N. 1979 Silage and milk production: a comparison between soya, groundnut and single-cell protein as silage supplements. *Grass and Forage Science* **34** 101-106

CASTLE, M.E., RETTER, W.C. & METCALFE, J.D. 1977a A note on supplements for dairy cows offered silage of high digestibility. *Animal Production* **25** 397-400

CASTLE, M.E., RETTER, W.C. & WATSON, J.N. 1977b Silage and milk production: a comparison between additives for silage of high digestibility. *Journal of the British Grassland Society* **32** 157-164

CASTLE, M.E., RETTER, W.C., WATSON, J.N. & ZEWDIE, E. 1977c Silage and milk production: a comparison between four rates of groundnut cake supplementation of silage of high digestibility. *Journal of the British Grassland Society* **32** 43-48

CASTLE, M.E., RETTER, W.C. & WATSON, J.N. 1979 Silage and milk production: comparisons between grass silage of three different chop lengths. *Grass and Forage Science* **34** 293-301

CASTLE, M.E., RETTER, W.C. & WATSON, J.N. 1980 Silage and milk production: a comparison between three grass silages of different digestibilities. *Grass and Forage Science* **35** 219-225

DULPHY, J.P. & DEMARQUILLY, C. 1975 [Influence of the type of forage harvester on the silage intake level and the performances of dairy cows.] *Annales de Zootechnie* **24** 363-371

ETTALA, ELSI & LAMPILA, M. 1978 Factors affecting voluntary silage intake by dairy cows. *Annales Agriculturae Fenniae* **17** 163-174

FARHAN, S.M.A. & THOMAS, P.C. 1978 The effect of partial neutralization of formic acid silages with sodium bicarbonate on their voluntary intake by cattle and sheep. *Journal of the British Grassland Society* **33** 151-158

FLYNN, A.V. 1979 Conservation of forage for beef cattle. *Proceedings, Grange Cattle Production Seminar, Paper No.* 8 Agricultural Institute, Grange, Co. Meath

FORBES, T.J. & JACKSON, N. 1971 A study of the utilization of silages of different dry-matter content by young beef cattle with or without supplementary barley. *Journal of the British Grassland Society* **26** 257-264

GORDON, F.J. 1975 Milk production from silage and concentrates containing varying proportions of dried grass and cereals. *Animal Production* **20** 173-179

GORDON, F.J. 1979a The effect of protein content of the supplement for dairy cows with access *ad libitum* to high digestibility, wilted grass silage. *Animal Production* **28** 183-189

GORDON, F.J. 1979b Management of the spring-calving dairy herd. *Occasional Publication No.* 4 Agricultural Research Institute of Northern Ireland, Hillsborough, Co. Down

GORDON, F.J. 1980 The effect of level of protein in the supplement given to dairy cows with *ad libitum* access to grass silage. *Forage Conservation in the 80's* (Ed. C. Thomas) *British Grassland Society, Occasional Symposium No.* 11 pp 428-431

GORDON, F.J. & McMURRAY, C.H. 1979 The optimum level of protein in the supplement for dairy cows with access to grass silage. *Animal Production* **29** 283-291

GORDON, F.J. & MURDOCH, J.C. 1978 An evaluation of a high-quality grass silage for milk production. *Journal of the British Grassland Society* **33** 5-11

LAIRD, R., LEGGATE, A.T. & CASTLE, M.E. 1979 The effect of supplementary protein on the performance of dairy cows offered grass silage *ad libitum. Animal Production* **29** 151-156

LAIRD, R., LEAVER, J.D., MOISEY, F.R. & CASTLE, M.E. 1981 The effects of concentrate supplements on the performance of dairy cows offered grass silage *ad libitum. Animal Production* **33** 199-209

LEAVER, J.D. 1980 Systems of feeding dairy cows. (a) Individual cow feeding systems. *Feeding Strategies for Dairy Cows* (Eds W.H. Broster, C.L. Johnson & J.C. Tayler) pp 10. 1-10. 11 Agricultural Research Council, London

McILMOYLE, W.A., MURDOCH, J.C. & GORDON, F.J. 1975 A note on dried grass as a component of concentrate mixtures for lactating dairy cows. *Animal Production* **20** 163-166

McLEOD, D.S., WILKINS, R.J. & RAYMOND, W.F. 1970 The voluntary intake by sheep and cattle of silages differing in free-acid content. *Journal of Agricultural Science* **75** 311-319

MARSH, R. 1978 A review of the effects of mechanical treatment of forages on fermentation in the silo and on the feeding values of the silages. *New Zealand Journal of Experimental Agriculture* **6** 271-278

MARSH, R. 1979 The effects of wilting on fermentation in the silo and on the nutritive value of silage. *Grass and Forage Science* **34** 1-9

MINISTRY OF AGRICULTURE, FISHERIES AND FOOD, DEPARTMENT OF AGRICULTURE AND FISHERIES FOR SCOTLAND & DEPARTMENT OF AGRICULTURE FOR NORTHERN IRELAND 1975 Energy allowances and feeding systems for ruminants. *Technical Bulletin* 33 Her Majesty's Stationery Office, London

MURDOCH, J.C. 1965 The effect of length of silage on its voluntary intake by cattle. *Journal of the British Grassland Society* **20** 54-58

ØSTERGAARD, V. 1979 Strategies for concentrate feeding to attain optimum feeding level in high yielding dairy cows. *Beretning fra Statens Husdyrbrugs forsog* 482 National Institute of Animal Science, Copenhagen

RETTER, W.C. 1978 The production, and the utilization of high-quality grass silage by dairy cows. *Glasgow University, Ph.D. thesis*

STEEN, R.W.J. & GORDON, F.J. 1980a The effect of level and system of concentrate allocation to January/February calving cows on total lactation performance. *Animal Production* **30** 39-51

STEEN, R.W.J. & GORDON, F.J. 1980b The effect of type of silage and level of concentrate supplementation offered during early lactation on total lactation performance of January/February calving cows. *Animal Production* **30** 341-354

TAYLER, J.C. & ASTON, K. 1973 Dried grass v. barley as a concentrate for milk production. *Grass — Journal of the British Association of Green Crop Driers* No. 6 pp 3-8

TAYLOR, K. 1979 Flat-rate feeding of concentrates to dairy cows. *Farm Management Services, Information Unit, Report No.* 20 Milk Marketing Board, Reading

THOMAS, P.C. & CASTLE, M.E. 1979 The work of the Nutrition and Metabolism and Dairy Husbandry Sections of the Applied Studies Department. *Hannah Research Institute Report* 1978 pp 108-117

THOMAS, C., ASTON, K. & DALEY, S.R. 1978 The potential of ensiled grass and forage crops for milk production by the autumn-calved cow. *Leaflet* Grassland Research Institute, Hurley, Maidenhead

WILKINS, R.J., HUTCHINSON, K.J., WILSON, R.F. & HARRIS, C.E. 1971 The voluntary intake of silage by sheep. 1. Interrelationships between silage composition and intake. *Journal of Agricultural Science* **77** 531-537

WILKINS, R.J., FENLON, J.S., COOK, J.E. & WILSON, R.F. 1978 A further analysis of relationships between silage composition and voluntary intake by sheep. *Proceedings, 5th Silage Conference* (Ed. R.D. Harkess) pp 34-35 Hannah Research Institute, Ayr

## Chapter 7

# Present conclusions and future research

**P.C. Thomas, M.E. Castle & J.A.F. Rook†**

*Hannah Research Institute, Ayr KA6 5HL*

It has been demonstrated that it is technically feasible to produce consistently, on a farm scale, silages containing approximately 11 MJ metabolizable energy per kg dry matter (DM) and more than 150 g crude protein per kg DM. These silages have a high feeding value and are especially useful in the achievement of satisfactory milk yields in high-forage systems of milk production or in the feeding of high-yielding cows, where the demands for energy and limitations on appetite preclude the feeding of low-energy forages. Notwithstanding this, however, there are areas of uncertainty, both in the production and use of high-quality silages, which require further study. Some unanswered questions are of a technical nature, and may be investigated in the laboratory and in controlled feeding experiments, but others are questions of farm management and economics, and will require solution through development work and experience on farms.

## HERBAGE PRODUCTION

Key features in any silage system are the type of sward and its management. The choice of the species and variety of grass will be influenced by agronomic and climatic considerations but in the western and northern parts of the United Kingdom there are strong arguments for a late perennial ryegrass, which has a satisfactory yield, a high digestibility (which declines relatively slowly with increasing maturity) and good characteristics for ensilage. Fertilizer and cutting policies must be integrated in the management of the farm as a whole, and may be influenced by the requirements of the 'non-silage' sector of the enterprise, e.g. pasture for grazing. However, there is a clear need to ensure that fertilizer applications are consistent with a high production of grass during the early part of the growing season and that the grass is cut before its digestibility declines to unacceptable levels. Inevitably, there will be some conflict between the attainment of high yields of grass DM

† Present address: Agricultural Research Council, 160 Great Portland Street, London W1N 6DT

and high digestibility of organic matter in the dry matter (DOMD); maximum yields of digestible organic matter (DOM) are probably achieved when DOMD is approximately 650 g. $kg^{-1}$ and cutting at higher concentrations leads to a reduction in nutrient production per ha. However, this is a subject for further research, since the yield of the first cut has a dominant influence on the total season's yield, and the timing and rate of application of nitrogen fertilizer for the first cut, and the time of cutting, are crucial. On most British farms there is scope for substantial improvement in grassland productivity and, under these circumstances, a slight loss in DOM yield per ha as a result of early cutting can be offset by higher rates of fertilizer application. Some development studies on systems of cutting grass for silage have been undertaken at the Crichton Royal Farm, Dumfries (Moisey & Leaver, 1980) and these indicate that a three-cut system designed to produce silage of high-digestibility leads to a loss of DOM yield per ha of approximately 8% (table 1). However, in this study the swards on the two-cut and three-cut systems were given equal amounts of fertilizer, and different results might have been obtained if the fertilizer treatments for each system had been optimum for the selected frequency of cutting.

A matter of some concern for the future is the high and increasing cost of fertilizers, which is central to the economics of grass production for silage. There appears to be scope for reducing fertilizer-N requirements through the use of mixed swards containing grass and one of the new strains of large-leaved white clover such as Blanca (RvP). There is, however, little information at present on the yields, composition and management of such swards for silage, and these subjects must be priorities for research. Other legumes, such as red clover, are useful silage crops in specific circumstances but are not likely to have widespread application.

## MAKING SILAGE

The principles fundamental to the production of high-quality silage are now well recognized and, provided they are adhered to closely, satisfactory products can undoubtedly be made using numerous variants of ensilage procedure. The system described on page 122 is simple and well tried, and at the Hannah Research Institute has given consistently good results under a range of conditions. Difficulties were experienced only in 1976 when the harvesting period was cool and wet, and the silage had a lower quality than intended. There would be advantages in the use of methods either to predict or to measure the chemical

**Table 1. The yields and digestible organic matter in the dry matter of grass silage produced by a three-cut and a two-cut system of grassland management (Moisey & Leaver, 1980)**

| | Three-cut system | | | Two-cut system | |
|---|---|---|---|---|---|
| | **1st** | **2nd** | **3rd** | **1st** | **2nd** |
| Dates of cutting | 22 May | 3 July | 15 + 16 Aug | 8 + 9 June | 15 + 16 Aug |
| Fertilizer (kg. ha$^{-1}$) $N:P_2O_5:K_2O$ | 150:0:0 | 75:37:37 | 75:37:37 | 150:0:0 | 150:75:75 |
| Dry matter (DM) yield (t. ha$^{-1}$) | 4.4 | 2.8 | 2.2 | 7.0 | 4.3 |
| Total DM yield (t. ha$^{-1}$) | | 9.4 | | 11.3 | |
| Digestible organic matter (DOM) in the DM (g. kg$^{-1}$) | 716 | 671 | 663 | 637 | 595 |
| DOM yield (t. ha$^{-1}$)† | 3.15 | 1.88 | 1.46 | 4.46 | 2.56 |
| Total DOM yield (t. ha$^{-1}$) | | 6.49 | | 7.02 | |

† Ensilage losses were approximately 180 g. kg$^{-1}$ on all treatments

composition of the grass prior to cutting, but for this to be feasible on a farm scale the methods would need to be simple and easily applicable.

A part of the silage-making procedure that may be particularly problematic is the wilting period. Events during this period are largely beyond the farmer's control, and poor weather has adverse effects on both ensilage losses and the quality of the silage produced. These difficulties may be avoided by ensiling unwilted grass directly after cutting and by using an effective additive and, if a satisfactory preservation is achieved, there is no objection to this method from a nutritional standpoint. In practice, however, direct-cut herbage of low DM concentration is more difficult to ensile than wilted herbage and is often associated with an unacceptably high effluent loss. These problems are not insurmountable but further research and development work in this area is needed.

Consideration should also be given to the possible use of additives designed to influence events within the silo to produce silage of specified nutritional characteristics. Additives to manipulate silage fermentation are a recognizable possibility and there may be ultimately scope for adding chemicals to improve digestibility; the latter type of additive has the special attraction that it would allow grass to be cut at a more mature stage of growth, with a consequent gain in DM yield per ha. The addition of supplementary foods at ensilage is a further prospective development which might be advantageous and, for example, high-energy diets have been made recently at the HRI by incorporating fats into the grass during silage making.

Great improvements in silage machinery have occurred over the past two decades, and there are managerial and nutritional advantages to be gained from the sophisticated precision-chop harvesters that are now in widespread use. Short-chopping grass during silage making is clearly beneficial but any future need for low-cost systems of milk production may militate against the high cost of complicated types of equipment At a research level there is a need for a renewed examination of silage making with simple, low-cost equipment.

## FEEDING SILAGE

The work that has been undertaken with high-quality, high-digestibility silage leaves little doubt about its potential as a food for dairy cows or, indeed, for rapidly growing stock. Even within the constraints of present techniques, there is clearly considerable scope on

**Table 2. Milk production by cows given silages of high or low digestibility *ad lib.*, with high or low rates of concentrate supplementation (Moisey & Leaver, 1980)**

| | High-digestibility silage† | | Low-digestibility silage‡ | | |
|---|---|---|---|---|---|
| | High concentrate | Low concentrate | High concentrate | Low concentrate | s.e. of difference |
| Silage intake (kg DM per d) | 8.62 | 10.83 | 8.47 | 10.30 | |
| Concentrate intake (kg DM per d) | 8.32 | 4.19 | 8.34 | 4.23 | |
| Milk yield (kg. d$^{-1}$) | 21.6 | 19.6 | 19.5 | 16.2 | 0.83*** |
| Fat (g. kg$^{-1}$) | 41.0 | 42.4 | 42.1 | 44.6 | 1.0 * |
| Protein (g. kg$^{-1}$) | 34.9 | 33.3 | 34.2 | 32.4 | 0.6 * |
| Lactose (g. kg$^{-1}$) | 47.7 | 47.5 | 48.0 | 46.3 | 0.4 ** |
| Body weight increase (kg. d$^{-1}$) | 0.47 | 0.32 | 0.40 | 0.20 | 0.06* |

† Silage from three-cut system (see table 1).
‡ Silage from two-cut system (see table 1).

most farms for reducing concentrate use and substituting high-digestibility silage. An example of this can be seen in table 2, where silages of high and low digestibility from the three-cut and two-cut systems of grass production respectively, described above, were given to autumn-calving Friesian cows from November to April and were supplemented with either high or low levels of dairy concentrates (Moisey & Leaver, 1980). As might be anticipated, the highest milk yields were obtained with the high-digestibility silage and the high rate of concentrate feeding. However, it is important to note that similar milk yields were obtained from cows given either the high-digestibility silage and the low concentrate rate or the low-digestibility silage and the high concentrate rate although, naturally, the former cows ate more silage.

The voluntary intake of silage is clearly a central issue in any future development of high-forage systems for milk production, and the factors regulating the intake of silage, and those determining the interaction effects between silage and other foods, are matters of great importance. The studies conducted so far have shown the importance of silage digestibility and fermentation quality in determining intake, and recent work has illustrated the scope for enhancing intake through an appropriate selection of supplementary foods. There is, however, an urgent need for a better understanding of the mechanisms involved so that a systematic approach to increasing the intake of silage can be made. Identification of the silage fermentation products that have an adverse effect on intake would be helpful, as would a better understanding of the effects of dietary starch, sugars and proteins on the breakdown of plant cell walls in the rumen.

There is also a need for more information on the nutritional factors influencing the utilization of silage energy and protein. In particular, there is a lack of understanding about the utilization of N by the rumen microbes, and about the factors regulating the amount and composition of the mixture of amino acids passing from the rumen to the small intestine. Equally, the significance of the low concentrations of methionine and lysine in the protein in the duodenum of silage-fed animals needs to be explored.

Since 1980 in England and Wales the payment scheme for milk has included separate components for the milk fat and solids-not-fat concentrations, and it seems likely that, in future, similar schemes, probably based on fat, protein and lactose, will operate throughout the UK marketing sector. As a consequence, milk producers will be concerned to formulate diets not only adequate nutritionally in energy and protein but also providing the mixture of nutrients that will maximize the production of individual milk constituents. This, ultimately, is a question of achieving an appropriate balance in the end-products of digestion absorbed from the gut; a better appreciation of the dietary factors affecting this mixture, and of the relationship between the end-products of digestion and the production of individual milk constituents, is essential if satisfactory methods of diet formulation are to be devised. Silage diets have milk fat enhancing qualities although the fat produced has a characteristic fatty acid composition, which may not be especially suitable for all types of manufactured milk products. Rations containing a high proportion of silage are also sometimes associated with low milk protein contents and, if milk is to be used for processing, this is a disadvantage. There is a need for an improved understanding of the factors that affect milk protein, to provide a basis for the production of silages and the formulation of silage diets designed to increase milk protein content.

Similarly, if high-silage, low-concentrate diets are offered to cows in early lactation, the limitations on silage intake may lead to energy undernutrition, and hence to excessive tissue mobilization and weight loss. The animals may therefore be short of the glucose needed to sustain normal metabolism. It may be possible to avoid this undernutrition by using high-energy, high-protein, low-bulk supplements and also to devise supplements that provide additional glucose in the small intestine, or avoid the adverse effects of starch fermentation in the rumen on silage intake. These approaches, which are currently under investigation, would give increased flexibility in silage use and provide a new option in the choice of feeding systems, which might allow milk producers to cope more readily with any future adverse changes in the economics of milk production.

Even at present, however, there is considerable scope on the average British farm for an improvement in the amount and quality of silage made for winter feeding, and for many milk producers there would be economic benefits from a reduced use of purchased concentrate foods and a fuller exploitation of the potential of home-grown high-quality silage.

## REFERENCE

MOISEY, F.R. & LEAVER, J.D. 1980 A comparison of three- and two-cut silage systems – II. *Crichton Royal Farm Report* 1979 pp 13-15 Advisory and Development Service, West of Scotland Agricultural College, Ayr

# Index